AF895017

MATHEMATICS RESEARCH DEVELOPMENTS

FOCUS ON NONLINEAR ANALYSIS RESEARCH

Mathematics Research Developments

Additional books in this series can be found on Nova's website
under the Series tab.

Additional e-books in this series can be found on Nova's website
under the e-book tab.

MATHEMATICS RESEARCH DEVELOPMENTS

FOCUS ON NONLINEAR ANALYSIS RESEARCH

GUALTIERO PADOVANI
AND
MANUEL OCCHINO
EDITORS

nova publishers
New York

Copyright © 2013 by Nova Science Publishers, Inc.

All rights reserved. No part of this book may be reproduced, stored in a retrieval system or transmitted in any form or by any means: electronic, electrostatic, magnetic, tape, mechanical photocopying, recording or otherwise without the written permission of the Publisher.

For permission to use material from this book please contact us:
Telephone 631-231-7269; Fax 631-231-8175
Web Site: http://www.novapublishers.com

NOTICE TO THE READER

The Publisher has taken reasonable care in the preparation of this book, but makes no expressed or implied warranty of any kind and assumes no responsibility for any errors or omissions. No liability is assumed for incidental or consequential damages in connection with or arising out of information contained in this book. The Publisher shall not be liable for any special, consequential, or exemplary damages resulting, in whole or in part, from the readers' use of, or reliance upon, this material. Any parts of this book based on government reports are so indicated and copyright is claimed for those parts to the extent applicable to compilations of such works.

Independent verification should be sought for any data, advice or recommendations contained in this book. In addition, no responsibility is assumed by the publisher for any injury and/or damage to persons or property arising from any methods, products, instructions, ideas or otherwise contained in this publication.

This publication is designed to provide accurate and authoritative information with regard to the subject matter covered herein. It is sold with the clear understanding that the Publisher is not engaged in rendering legal or any other professional services. If legal or any other expert assistance is required, the services of a competent person should be sought. FROM A DECLARATION OF PARTICIPANTS JOINTLY ADOPTED BY A COMMITTEE OF THE AMERICAN BAR ASSOCIATION AND A COMMITTEE OF PUBLISHERS.

Additional color graphics may be available in the e-book version of this book.

LIBRARY OF CONGRESS CATALOGING-IN-PUBLICATION DATA

ISBN: 978-1-62417-350-9

Published by Nova Science Publishers, Inc. † New York

CONTENTS

PREFACE

In this book, the authors present current research on nonlinear analysis, from across the globe. Topics include modelling of the characteristics of inelastic members of reinforced concrete structures in seismic nonlinear analysis; automatic estimation of the apnea-hypopnea index from oxygen saturation recordings using non-linear regression analysis; nonlinearity in bridge structural analysis; nonlinear analysis of spontaneous MEG activity in Alzheimer's disease; mesh free-enriched finite element methods for nonlinear analysis of path-independent and path-dependent problems in solids; nonlinear analysis research in biomedical engineering; and nonlinear evolution analysis for flow during alloy crystallization.

Chapter 1 - In this Chapter, the issue of the modelling of the nonlinear characteristics of structural r/c elements is fully presented, where all the necessary information according to Eurocode EN 1998-3 in combination with the newly Hellenic Code of Retrofitting of Reinforced Concrete Buildings (KANEPE 2012) are given in detail. The point of the available chord rotations at yield and ultimate state of the structural r/c members (columns, beams, walls) and cores is analytically developed. Moreover, the point of the available cyclic shear strength of r/c members and the issue of the effective flexural stiffness of r/c member sections are given.

The natural meaning with the calculation way of the final Moment-Chord Rotation of r/c members is presented. In addition, the practical rational modelling of the masonry infill walls for the daily needs of r/c building design is described. Furthermore, the calculation of the seismic demand chord rotations (elastic ones and plastic ones) of the structural r/c members without using various springs at their ends is analytically presented.

Finally, a numerical example of an r/c cantilever with nonlinear behavior is analytically examined for illustrative reasons, since the cantilever always is the theoretical base for the complete modelling of each r/c building according to both above mentioned contemporary seismic Codes.

Chapter 2 - Generalized regression neural networks (GRNN) represent a powerful tool for modeling non-linear relationships between variables. In this study, a GRNN-based model is developed to assist in the diagnosis of sleep apnea-hypopnea syndrome (SAHS) from nocturnal oxygen saturation (SaO_2) recordings. SAHS has been pointed out as a major cause of traffic and industrial accidents and long-term effects can lead to severe cardiovascular diseases. Nowadays, nocturnal polysomnography (PSG) is the gold-standard for SAHS diagnosis. PSG must be performed in a special sleep unit and under supervision of a trained technician. Different physiological signals recorded during a complete night must be analyzed by an expert to obtain the apnea-hypopnea index (AHI). It measures the ratio of apneas and hypopneas per hour of sleep and determines SAHS severity. Despite its high diagnostic performance, PSG has several drawbacks since it is complex, costly and time-consuming. Thus, simplified diagnostic techniques are desirable. This study proposes analyzing SaO_2 signals recorded through nocturnal pulse oximetry to assist in SAHS diagnosis. The aim is to obtain a reliable estimation of the AHI from automatic analysis of these recordings. A pattern recognition approach based on GRNN was used to implement the authors'diagnostic method. The proposed methodology comprises three different stages: 1) signal preprocessing, 2) feature extraction and 3) pattern analysis. In the first one, signal artifacts are removed. Subsequently, a total of 14 time-domain and frequency-domain features were extracted from SaO_2 recordings. In the third stage, these features were used as inputs to a GRNN model in order to estimate the AHI. A database composed of 240 SaO_2 signals was available for this study. It was randomly divided into a training set (40%) with 96 SaO_2 signals and a test set (60%) with 144 SaO_2 signals. The authors found that their GRNN-based model achieved an intraclass correlation coefficient (ICC) of 0.89 on the test set, showing high reliability in the estimation of the AHI. According to theirresults, the proposed GRNN model represents a valuable screening tool for SAHS. It could be used in clinical practice in order to reduce the number of required PSG tests.

Chapter 3 - The term “nonlinear system” refers to a given structure undergoing a change in stiffness in its loaded state. For linear analyses, intermediate steps between stages are not required. However, this is a necessity when dealing with nonlinear problems.

There are, in general, two types of nonlinearity. The first type is called “material nonlinearity”, in which the nature of the material changes the stiffness of the structure as it is loaded. The second type is called “geometric nonlinearity”, in which a given structure undergoes large displacements relative to the size of the structure. This causes large changes in the geometric stiffness of the structure. Material nonlinearity covers the subjects of the stress-strain relationships corresponding to unconfined concrete, confined concrete, and longitudinal steel reinforcement and others, such as Bauschinger effect, stress relaxation and creep. Geometric nonlinearity includes second-order effects on a structure, as well as instability of a large structure. The original deflection theory and modern matrix analyses with geometric nonlinearities for long-span bridges are discussed, respectively. Other types of nonlinearity such as nonlinearity due to change of status are also covered. In order to demonstrate the usage, nonlinear analyses of long-span bridges (cable stayed bridges and suspension bridges) and nonlinear seismic analyses of bridge structures (nonlinear time history analysis and modal pushover analysis with their comparison) are also given in this chapter.

Nowadays, with the rapid development of modern day software, many sophisticated structural programs can be used for solving nonlinear problems under extreme limit states to complete the bridge design and checking process. With these tools, nonlinear problem solving becomes more or less a routine process for structural engineers. Through this article, engineers may become familiar with those complexities and assumptions to analyze bridges in the nonlinear range to make them safe and economical.

Chapter 4 - Alzheimer's disease (AD) is one of the most common disorders among elderly population and it is considered the main cause of dementia in western countries. This progressive, degenerative brain disorder is characterized by neural loss and the appearance of neurofibrillary tangles and senile plaques. Although a definite diagnosis is only possible by necropsy, a differential diagnosis with other types of dementia and with major depression should be attempted. The differential diagnosis includes medical history studies, physical and neurological evaluation, mental status tests, and neuroimaging techniques. Nowadays, magnetoencephalography (MEG) recordings are not used in AD clinical diagnosis, despite its potential as aid diagnostic tool.

The aim of this chapter is to show recent analyses of the spontaneous MEG activity in AD. For this purpose, five minutes of recording were acquired with a 148-channel whole-head magnetometer in 36 patients with probable AD and 26 control subjects. MEG data were analyzed by means of three nonlinear measures: Lempel-Ziv complexity (*LZC*), sample entropy (*SampEn*) and detrended fluctuation analysis (*DFA*). *LZC* measures the number of different substrings and the rate of their recurrence along the original time series. *SampEn* is an embbeding entropy that quantifies the signal regularity. Finally, *DFA* is a method designed to quantify correlations in noisy and non-stationary time series.

LZC and *SampEn* showed that MEG recordings are less complex and more regular in AD patients than in control subjects. *DFA* revealed changes in the fluctuations of MEG signals. Significant differences between AD patients and elderly controls were found using the three nonlinear methods (p-values < 0.05, Welch t-test). The authors used receiver operating characteristic (ROC) curves with a leave-one-out cross-validation procedure to assess the ability of the methods to classify AD patients and control subjects. The highest area under the ROC curve was achieved with *DFA* (0.7959) whereas the highest accuracy was obtained with *LZC* algorithm (72.58%).

The authors conclude that nonlinear analyses from spontaneous MEG activity could be useful to increase their insight into AD. This chapter shows that MEG recordings reflect the alterations on AD patients' brains. These alterations may be associated with deficiencies in information processing.

Chapter 5 - This chapter presents a large deformation analysis of nonlinear path-independent and path-dependent problems based on Meshfree-enriched Finite element Method (ME-FEM). In ME-FEM, the problem domain in two-dimension is first discretized by a regular triangulation using linear triangular elements. The element formulation is then established by introducing a meshfree convex approximation into the linear triangular element along with an enriched meshfree node. To provide a volumetric locking-free analysis for near-incompressible materials, an area-weighted smoothing scheme for deformation gradient is developed in conjunction with the meshfree-enriched element interpolation functions to yield a discrete divergence-free property at the integration points for both path-independent and path-dependent problems. A three-field Hu-Washizu variational functional is considered for the development of nonlinear finite element formulation. Since the meshfree-enriched element shape functions are constructed using the meshfree convex approximation, they pose the desired Kronecker-delta property at the element edge thus requires no special treatments in the enforcement of essential boundary condition as well as the contact conditions. The variational formulation is further degenerated to become a primal problem based on the proposed area-weighted smoothing scheme for deformation gradient. As the results, the proposed method can be easily implemented into a conventional displacement-based finite element code. Several hyper-elasticity and elasto-plasticity problems are studied to demonstrate the effectiveness of the proposed method. The numerical results indicated that the proposed method is capable of delivering a volumetric locking-free and pressure oscillation-free solution solutions to near-incompressible materials in large deformation.

Chapter 6 - Biological phenomena have shown behaviors that are strongly nonlinear. Therefore, in order to correctly enhance information from biomedical signals, analysis based on nonlinear methods should be adopted beyond the classical linear modeling. These approaches reflect better the nonlinear nature of biological phenomena, and are more suitable to model and interpret information of modulation phenomena, on-off mechanism of response, variable gain, threshold and saturation characteristics, and even chaotic behaviors. In this chapter, the authors will review nonlinear methods applied for the processing of biomedical signals. The authors will refer to the Heart Rate Variability, the series of heartbeats, as this signal is commonly analyzed using both linear and nonlinear methods. Nonetheless, these methods have been widely applied to the study of other biomedical signals. The methods reviewed in this chapter are the Poincaré plot, the fractal dimension, the correlation dimension, the detrended fluctuation analysis, the approximate entropy, the sample entropy, and the recurrence plot. The authors will show the potentiality of nonlinear methods compared to the linear modeling with reference to published studied.

Chapter 7 - This article presents an analysis of nonlinear evolution equation appearing in fluid flow during solidification of a binary alloy which is cooled from below. It has been well established by experimentalists that a horizontal mushy layer is formed during the solidification of binary alloys. The authors have included a review section on recent studies done by various researchers on this topic. This chapter includes an analysis on nonlinear behavior of the convective flow for a passive as well as an active mushy layer. The effects of presence and absence of a magnetic field in the mushy layer are also considered.

The basic state, linear, adjoint and first-order systems are derived and these solutions are used to obtain the Landau coefficients which appear in the resulting nonlinear evolution equation. The authors' computational results suggest that higher permeability parameter and Chandrasekhar number enhance the marginal stability. This analysis also indicates a possible slow transition to turbulence in such flow system since the computed Landau constants are positive.

In: Focus on Nonlinear Analysis Research
Editors: Gualtiero Padovani and Manuel Occhino
ISBN: 978-1-62417-350-9
© 2013 Nova Science Publishers, Inc.

Chapter 1

MODELLING OF CHARACTERISTICS OF INELASTIC MEMBER OF REINFORCED CONCRETE STRUCTURES IN SEISMIC NONLINEAR ANALYSIS

Triantafyllos K. Makarios*

Earthquake Engineering Division, Institute of Engineering Seismology and Earthquake Engineering, Thessaloniki, Greece

ABSTRACT

In the present Chapter, the issue of the modelling of the nonlinear characteristics of structural r/c elements is fully presented, where all the necessary information according to Eurocode EN 1998-3 in combination with the newly Hellenic Code of Retrofitting of Reinforced Concrete Buildings (KANEPE 2012) are given in detail. The point of the available chord rotations at yield and ultimate state of the structural r/c members (columns, beams, walls) and cores is analytically developed. Moreover, the point of the available cyclic shear strength of r/c members and the issue of the effective flexural stiffness of r/c member sections are given.

The natural meaning with the calculation way of the final Moment-Chord Rotation of r/c members is presented. In addition, the practical rational modelling of the masonry infill walls for the daily needs of r/c building design is described. Furthermore, the calculation of the seismic demand chord rotations (elastic ones and plastic ones) of the structural r/c members without using various springs at their ends is analytically presented.

Finally, a numerical example of an r/c cantilever with nonlinear behavior is analytically examined for illustrative reasons, since the cantilever always is the theoretical base for the complete modelling of each r/c building according to both above mentioned contemporary seismic Codes.

Keywords: Simulation of plastic hinge properties, masonry infill walls, seismic nonlinear static analysis, seismic nonlinear response history analysis

* 5 Ag. Georgiou Str., Patriarchika Pylaia, Thessaloniki, GR-55535. Tel.: (+)30 2310 476081, Fax: (+)30 2310 476085. Corresponding author, e-mail: makarios@itsak.gr.

1. INTRODUCTION

In order to apply the seismic nonlinear analysis (static or dynamic), the modelling of the seismic characteristics of inelastic members of reinforced concrete (r/c) structures must be rationally taken place, previously. These characteristics are dependent mainly from the plastic hinges that can be developed near of the two ends of each structural r/c element (beams, columns and at base of r/c walls or cores). General, with reference to abovementioned characteristics, the following most important issues must be known:

(a) the available Moment-Chord Rotation (M - θ) diagram at each critical end of each structural member,
(b) the available cyclic shear strength of each structural r/c member, since it clarifies if the flexural (ductile) failure precedes than the shear (brittle) one,
(c) the effective flexural stiffness of r/c member sections and
(d) the practical modelling of the masonry infill walls

In addition, the issue of the calculation of seismic demand chord rotations (elastic part and plastic one) of the structural r/c members without use springs at their ends is always in timelines. In the past, many techniques about the simulation of the r/c structures in nonlinear seismic analysis presented, i.e. see Banon *et al* (1981) and at there references, but here, we are going to focus to contemporary Eurocode EN 1998-3 (2005) in combination with the newly Hellenic Code of Retrofitting of r/c Buildings (KANEPE, 2012). These two later Seismic Codes propose a flexural nonlinear law of Moment-Chord Rotation at each critical end-section of each r/c member of the structure. Equivalently, we can say that each structural r/c member is based on the seismic behavior of a cantilever for each critical end-section. It is known that, in order to estimate the seismic demands at low seismic performance levels according to Eurocode EN 1998-3, such as "*significant damage*" (that correspond with the "*life safety*" level) and "*near collapse*" (that correspond with the "*collapse prevention*" level), the application of the nonlinear response history analysis is suggested for accuracy reasons. On the contrary, due to high computational cost, civil engineers often prefer to use the simplified nonlinear static analysis (pushover analysis) and for this reason, the later analysis is proposed by EN 1998-3 and KANEPE 2012 as a basic nonlinear seismic method of seismic analysis.

In the case of the nonlinear response history analysis, there is a big space in the qualitative comprehension of the structure behavior, while there are various difficulties to verification of results. We are going back, in the case of the pushover analysis, there is natural inspection with smaller computational cost, issues that are both very important. From the international relative literature, a more rational procedure is that; the combination of the nonlinear static analysis with the inelastic response acceleration spectra (ATC 40, ATC-63, FEMA 273, FEMA P-440A, FEMA P-695). However, in order to combine the both abovementioned things, namely the pushover analysis and the response spectra, a suitable equivalent NonLinear Single Degree of Freedom (NLSDoF) system that represents the initial real structure has to be previously defined, because it is known that the response spectra are supported on a kind of SDoF system. Thus, the basic meaning of an SDoF system that can represent the initial structure came from the *Q*-model developed by Saiidi and Sozen (1981).

However, this idea has been gradually developed in more mature form by Fajfar and Fischinger (1987a,b). Other improved version of equivalent SDoF took space by Uang and Bertero (1990), Qi and Moehle (1991), Rodriquez (1994), Fajfar and Gaspersic (1996) and Makarios (2005).

Final, a more improved optimum equivalent NonLinear Single Degree of Freedom system, which is suitable for spatial (3D) asymmetric multi-storey buildings taking fully into account the floor rotations around vertical axes presented by Makarios (2009, 2010, 2012a). In these later works has been proved that; "the optimum NLSDoF of planar multi-storey frames can be supposed as a special case of the total one of spatial multi-storey building" and this fact is very important. Once an NLSDoF system represents, approximately, the real asymmetric multi-storey building, the seismic target-displacement of this system is calculated using the known procedure of inelastic response spectra (Krawinkler and Nassar, 1992; Chopra and Goel, 1999; Fajfar, 1999 and 2000). Afterwards, the seismic target-displacement of a NLSDoF system is transformed in seismic demand displacement at monitoring point of the real structure and from its known pushover analysis, the seismic demand analysis results of the structure can be obtained.

Moreover, the use of improved optimum equivalent NonLinear Single Degree of Freedom system can be suitable and in irregular in-plan, asymmetric, torsionally-flexible multi-stotey r/c buildings (Makarios, 2012a). It is worth to note that in the case of irregular in the plan buildings, according to sect.4.3.3.4.2.1(2)P of Eurocode EN 1998-1 and sect.4.4.4.1(2)P of Eurocode EN 1998-3, a suitable spatial model of the structure has to used in seismic nonlinear analysis.

However, in this case, no additional details have been given by EN 1998 about the spatial model of the structure to be used in conjunction with the pushover procedure described in this Eurocode with result the use of more accuracy method, such inelastic response history, to be obligatory. In addition, from a parametric analysis of such irregular multi-storey building, the suitability of this optimum NLSDoF has been verified (Makarios, 2009 and 2010). On the other hand, in seismic nonlinear response history analysis does not require the use of an equivalent Single Degree of Freedom system, while the seismic action is taken into account via suitable accelerograms at the base of the structure. However, based on previous experiences with nonlinear response history analyses, the results may be deemed unreliable due to the following reasons:

- Use of unsuitable accelerograms, where, inadequate number of recorded accelerograms, limited seismic data at the site, inadequate number of strong cycles of the dynamic loading in seismic excitation, inadequate strong motion duration as well as inadequate of the Arias Intensity (Arias, 1970) and the frequency content of used accelerograms.
- Inserting of false assumptions in the numerical models of the structures about the nonlinear dynamic properties of plastic hinges via the Moments-Chord Rotations (M - θ) diagrams of the structural members.
- Inadequacy relative to accuracy and stability of the numerical integration methods,
- Inadequacy in orientation of the pair of horizontal seismic components, since it is known that the critical dynamic loading orientation of the above pair is unknown or

does not exist; this issue leads to the examination of various other orientations in nonlinear response history analysis (Athanatopoulou, 2005).

- Omitting the vertical seismic ground motion component and ignoring the *P-Delta* effects in the analysis.

Consequently, taking into account all abovementioned things, the rational and right modelling of the nonlinear characteristics is a requirement for the both types of nonlinear analysis, the static, and the dynamic seismic analysis. This right modelling is the main target of this Chapter, where all the necessary information for the above issues is given in detail. Moreover, a numerical example of a typical r/c cantilever is presented, since seismic behavior of the cantilever is the theoretical base for the completely modelling of the r/c structure according to both abovementioned seismic Codes.

2. Available Chord Rotations of Structural R/C Members of Building

2.1. General

For the needs of the seismic nonlinear analysis of the structures, two techniques can be applied. The first one that uses elements with "*lumped nonlinearity/plasticity*" and the second one that uses elements with "*distributed nonlinearity/plasticity*". The first technique is preferred for the practical estimation of the building seismic behavior (static or dynamic), while the second technique that has high computational cost is used for research mainly reasons. All contemporary Seismic Codes or Directives, such as Eurocode EN 1998-3, KANEPE 2012, FEMA P-440A and FEMA P-695, propose the first technique to obtain the building seismic nonlinear results. In the past, many models with "*lumped plasticity*" proposed, starting from the serial model by Giberson (1967). However, after many attempts by researchers of all World, this concept has been gradually developed in more mature form by Kanaan and Powell (1975) and Litton (1975) in the well-known program DRAIN-2D, where suitable springs with nonlinear law of Moment-Chord Rotation (M-θ) have been serially inserted at the ends of the structural r/c elements. Keeping this form and in order to build a model of an r/c building, each member (column or beam) can be assumed that it has deformed anti-symmetrically, while the structural wall can develop plastic hinge at its base section only (Figure 1).

Thus, each deformed structural member may be considered to consists of two "cantilevers", each having a length L_S, which is known as "*shear length*". In the case of ductile r/c walls, where they have shear ratio $a_S > 2.50$, the shear length L_S is equal to the distance from the base of wall until the zero-moment point due to a temporary lateral static loading of the building, (Figure 1). In the case of columns or beams, their shear length L_S is equal to the distance from the end of the element until the zero-moment point, while, approximately, we can consider that shear length L_S is equal to one-half of the clear length of the examined structural element according to secti.A.3.2.4(5) of EN 1998-3.

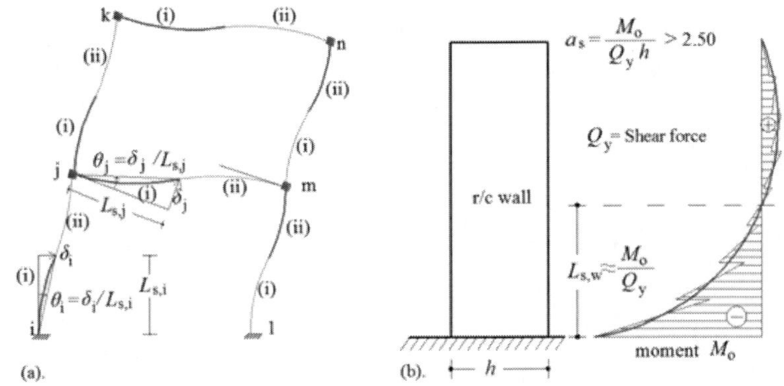

Figure 1. (a) Each element of an r/c frame consists of two sub-cantilevers (i) and (ii). (b) Definition of the shear length $L_{s,w}$ on r/c walls.

At the end-section of the base of each "cantilever", a suitable nonlinear spring is usually inserted in the building model, which follows a particular nonlinear law of Moment-Chord Rotation (M-θ). However, in order to obtain the nonlinear diagram M-θ, first, an elastoplastic (with 0% slope of the post-elastic part) diagram of Moment-Curvature (M - φ) has to be calculated for the base critical r/c section of each cantilever. This can be reliably achieved by local modelling of the final designed r/c section, applying the "fiber elements" (XTRACT, 2007). It is worthy to note that, the mean values of material strengths have to applied (i.e. $f_{cm} = f_{ck} + 8$ in MPa for concrete and $f_{ym} = 1.10 f_{yk}$ for steel) instead of their characteristic values f_{ck} and f_{yk} in order to calculate the reliable diagram of Moment-Curvature (M - φ). More details, the critical r/c section is divided into the field of confined concrete (which extends up to the loop of the internal line of the external stirrup), in field of unconfined concrete (which is outside of the loop of the internal line of the external stirrup) and in longitudinal steel bars of the section (Figure 2). For each one of the three fields abovementioned, a different appropriate stress-strain diagram (σ-ε) is used. However, on the one hand, the action of shear deformations and, on the other hand, the phenomenon of extraction or lap-splice slip of longitudinal steel bars from the fixed-base (i.e. join) of the cantilever are usually avoided and for this reason the exact analytic results of diagram Moment-Chord Rotation (M-θ) must be more corrected according to relative experimental data. However, at this issue we are going to come back later.

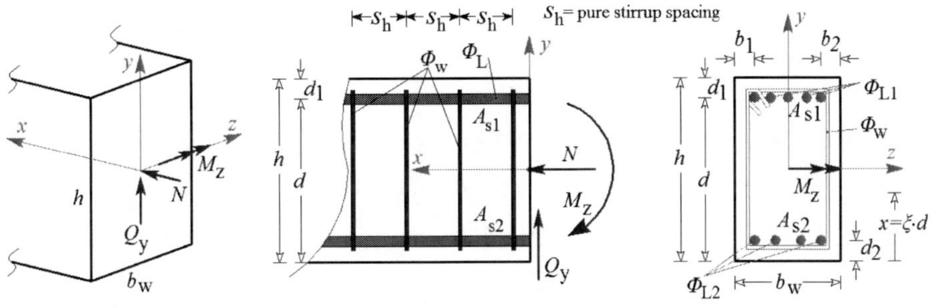

Figure 2. Section basic characteristics.

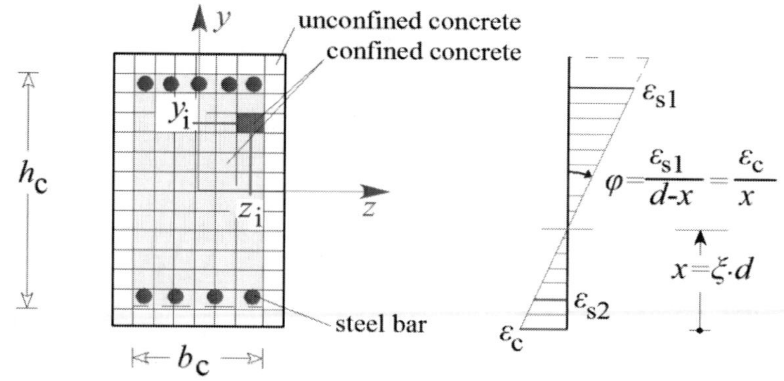

Figure 3. Section analysis using fiber elements and section strains along longitudinal x-axis.

2.2. Stress-Strain Diagram of Unconfined Concrete

According to paragraph 3.3 of Eurocode EN 1998-3, for choosing the admissible type of analysis and the appropriate Confidence Factor (CF) values of data of existing structures, three "*Knowledge Levels*" about the geometry of the structure, various details of reinforcement and the mechanical properties of materials, KL1 (limited knowledge), KL2 (normal knowledge) and KL3 (full knowledge), are defined. In addition, according to Table 3.1 of EN 1998-3, the recommended values of the Confidence Factors are 1.35, 1.20 and 1.00 for the KL1, KL2 and KL3, respectively. Thus, in order to calculate the Moment-Chord Rotation diagrams, the mean strength values of the materials must be divided with the abovementioned CF with regard to knowledge level of the data.

In the case of unconfined concrete, the analytical expression of its stress-strain σ - ε diagram is given by the known Hognestad parabola (Hognestad, 1951), where compressive strain shown positive:

$$\sigma_c = \frac{f_{cm}}{CF} \cdot \left(\frac{2 \cdot \varepsilon_c}{\varepsilon_{c2}} - \frac{\varepsilon_c^2}{\varepsilon_{c2}^2} \right) \quad \text{for } \varepsilon_c \le \varepsilon_{c2} = 0.002 \tag{1}$$

$$\sigma_c = \frac{f_{cm}}{CF} - \left(\frac{\left(f_{cm}/CF \right) - f_{cu}}{\varepsilon_{cu} - \varepsilon_{c2}} \right) \cdot \left(\varepsilon_c - \varepsilon_{c2} \right) \quad \text{for } \varepsilon_{c2} = 0.002 < \varepsilon_c \le \varepsilon_{cu} = 0.0035 \tag{2}$$

where $\varepsilon_{cu} = 0.0035$ and the respective strength is $f_{cu} = 0.75 f_{cm}/CF$.

In addition, the Elasticity Modulus E_c of the 28-days concrete is defined according to EN 1992-1-1 and EN 1998-3 as the slope of the line from the origin point to strength $0.40 f_{cm}/CF$ (Figure 4). Approximately, the Elasticity Modulus can be estimated as $E_c = 9.50 \cdot f_{cm}^{1/3}$, where E_c and f_{cm} in GPa. Alternatively, according to section 3.1.2(6) of Eurocode EN 1992-1-1, the compressive strength of concrete at an age t days depends on the type of cement, temperature, and curing conditions.

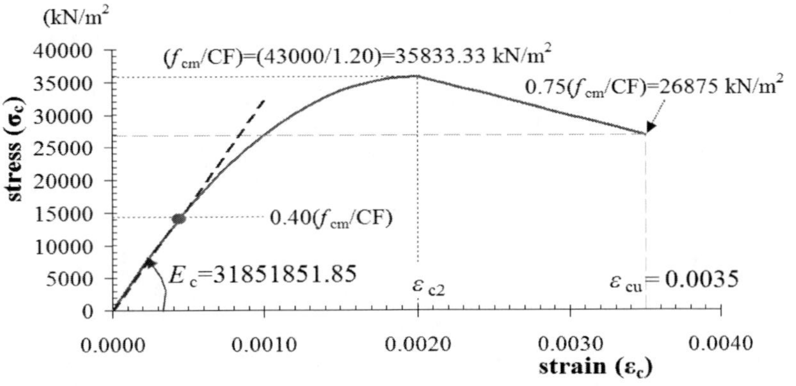

Figure 4. Stress-strain diagram σ - ε of unconfined concrete. Application for experimental in-situ mean strength $f_{cm} = 43\,\text{MPa}$ and Confidence Factor CF=1.20.

For a mean temperature of 20°C and curing in accordance with EN 12390 (European Norm, 2008), the compressive strength of concrete at various ages $f_{cm}(t)$ may be estimated from following expression:

$$f_{cm}(t) = f_{cm} \cdot e^{s \cdot \left(1 - \frac{5.292}{\sqrt{t}}\right)} \tag{3}$$

where, f_{cm} is the mean compressive strength at 28 days, t is the age of the concrete in days (after the first 28 days), s is a coefficient which depends on the type of cement (0.20 for fast coagulation, 0.25 for fast coagulation and high strength, 0.38 for normal).

2.3. Stress-Strain Diagram of Cold Worked Steel

According to Table C.1 of Annex C of Eurocode EN 1992-1-1, the characteristic strain of steel bars at maximum force is given $\varepsilon_{uk} = 0.075$ (for C steel class). However, according to section 3.2.7.(2) of EN 1992-1-1, the recommended value is the 90% of ε_{uk}, namely $\varepsilon_{uk} = 0.90 \times 0.075 = 0.0675$. Moreover, according to section C.2(1)P of Annex C (EN 1992-1-1), the maximum actual yield stress $f_{y,max}$ shall not exceed $1.30 f_{yk}$. In addition, the maximum failure stress is $f_u = k \cdot f_{yk}$, where factor k is taken by Table C.1 of Annex C (EN 1998-1-1) and for steel class C, k is from 1.10 to 1.35, while the design value of the Elasticity Modulus is taking $E_s = 200\,000.\text{MPa}$ (Figure 5). Idealized curve [A] of Figure 5 can be used for assessment of seismic capacity and retrofitting of existing buildings, but on the contrary, curve [B] is used for design. It is clear that, in the case where yield stress of steel bars is resulted by experimental data as it happened for assessment and retrofitting of existing buildings, the abovementioned Confidence Factors must be used. Thus, yield strain ε_y by mean yield stress f_{ym} (not characteristic f_{yk}) is given:

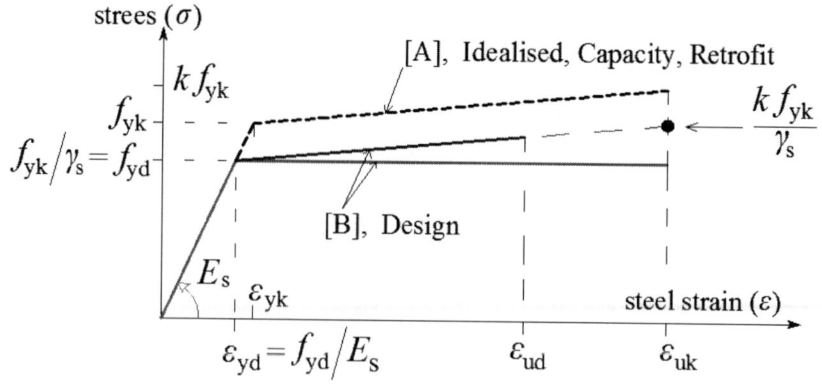

Figure 5. Stress-strain diagram σ - ε of cold worked steel with factor k=1.10-1.35 and Elasticity Modulus E_s=200GPa according to EN-1992-1-1.

$$\varepsilon_{\text{y}} = \frac{f_{\text{ym}}/\text{CF}}{E_{\text{s}}} \tag{4}$$

Afterwards of abovementioned, the analytical expression of stress-strain σ - ε diagram for steel bars is given, while in Figure 6 an application by experimental data with mean yield stress $f_{\text{ym}} = 545\,\text{MPa}$ and Confidence Factor CF=1.20 is shown:

$$\sigma_{\text{s}} = \varepsilon \cdot E_{\text{s}} \qquad \text{for } 0 \le \varepsilon \le \varepsilon_{\text{y}} = \frac{f_{\text{ym}}/\text{CF}}{E_{\text{s}}} \tag{5}$$

$$\sigma_{\text{c}} = \frac{f_{\text{ym}}}{\text{CF}} \cdot \left(1 + \frac{(k-1)(\varepsilon - \varepsilon_{\text{y}})}{\varepsilon_{\text{uk}} - \varepsilon_{\text{y}}} \right) \qquad \text{for } \varepsilon_{\text{y}} < \varepsilon \tag{6}$$

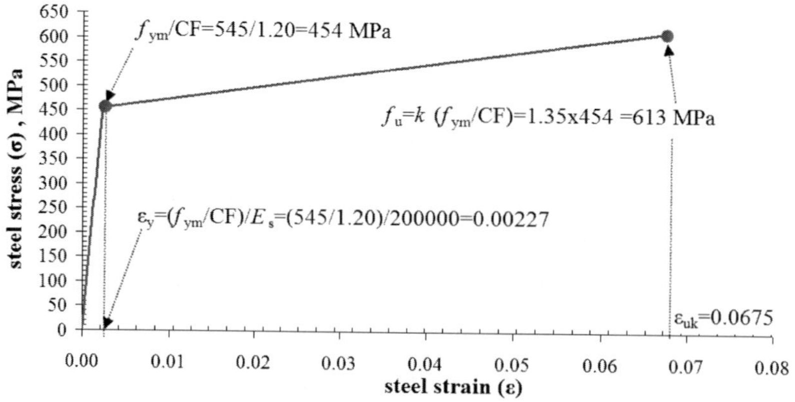

Figure 6. Stress-strain diagram σ - ε of cold worked steel according to EN-1992-1-1, using mean yield stress 545MPa, Confidence Factor CF=1.20, factor k=1.35 and Elasticity Modulus E_s=200GPa.

2.4. Stress-Strain Diagram of Confined Concrete

The stress-strain diagram σ-ε of confined core-concrete can be calculated based on the "*model of confined concrete*" that is proposed by Eq.(A.6-A.8)/sect. A.3.2.2 of Eurocode EN 1998-3. In that model, the strength f_{cc} of the confined concrete and its contemporary strain ε_{cc} is given as follows:

$$f_{cc} = \frac{f_{cm}}{CF}\left[1+3.7\cdot\left(\frac{a\cdot\rho_{sy}\cdot f_{yw,m}}{f_{cm}}\right)^{0.86}\right] \quad (7)$$

$$\varepsilon_{cc} = \varepsilon_{c2}\left[1+5\cdot\left(\frac{f_{cc}}{f_{cm}/CF}-1\right)\right] \quad (8)$$

where f_{cm} and ε_{c2} are the compressive strength (mean value) and corresponding strain of unconfined concrete, respectively (Figure 4).

The ultimate strain ε_{cu} of the extreme fiber of the pressure zone of the section is given:

$$\varepsilon_{cu} = 0.004+0.5\cdot\frac{\alpha\cdot\rho_{sy}\cdot\left(f_{yw,m}/CF\right)}{f_{cc}} \quad (9)$$

where $f_{yw,m}$ is the yielding stress (mean value) of the stirrups and α is the "*confinement effectiveness factor*" of the core of the section that is given as:

$$\alpha = \left(1-\frac{s_h}{2b_c}\right)\cdot\left(1-\frac{s_h}{2h_c}\right)\cdot\left(1-\frac{\sum b_i^2}{6\cdot b_c\cdot h_c}\right) \quad (10)$$

where b_i is the centerline spacing of longitudinal bars laterally restrained by a stirrup corner along the perimeter of the cross-section, so if b_i is less than 0.15m then the buckling phenomenon of these steel bars is eliminated. h_c and b_c is the dimension of confined core to the internal line of the external hoop.

It should be noted that, in the case when the stirrups are not closed with hooks that have an angle of more than 45°, then concrete confinement must be ignored and for this reason the "*confinement effectiveness factor*" is set to zero ($\alpha = 0$).

$\rho_{sy} = \dfrac{A_{sw,y}}{b_w\cdot s_h}$ is the ratio of transverse steel parallel to the loading local direction *y* of the section (Figure 2), $A_{sw,y}$ is the total area of the stirrup sections along the loading direction *y* and s_h is the pure stirrup spacing along the length *x* of the structural member.

After those, the analytical expression of Stress-Strain diagram σ-ε of confined core-concrete is given below:

$$\sigma_c = f_{cc} \cdot \left(\frac{2 \cdot \varepsilon_c}{\varepsilon_{cc}} - \frac{\varepsilon_c^2}{\varepsilon_{cc}^2} \right) \qquad \text{for } \varepsilon_c \leq \varepsilon_{cc} \qquad (11)$$

$$\sigma_c = f_{cc} - \left(\frac{f_{cc} - f_{cu}}{\varepsilon_{cu} - \varepsilon_{cc}} \right) \cdot \left(\varepsilon_c - \varepsilon_{cc} \right) \qquad \text{for } \varepsilon_{cc} < \varepsilon_c \qquad (12)$$

The following example is presented for illustrative reasons.

2.4.1. Numerical Example

Consider the r/c cantilever column of Figure 7, which has a concentrated mass $m = 12.00\,\text{t}$ at its top. The column section with reinforcements is formed as we can see in the same figure, where there are 24 steel bars of section diameters 14mm, thus $A_{s,tot} = (1.54\,\text{cm}^2) \times 24 = 36.96\,\text{cm}^2$, and stirrups of section diameter 8mm located along the length of the column (x-axis) per mean spacing 8 cm (thus pure spacing $s_h = 6.4\,\text{cm}$). By in-situ experimental data are known the concrete mean strength is $f_{cm} = 43\,\text{MPa}$, the steel mean yield stress is $f_{ym} = 545\,\text{MPa}$ and the steel failure strength is $f_u = 735\,\text{MPa}$. Moreover, consider that the there is "normal knowledge", namely level KL2, so we are going to use Confidence Factor equal with 1.20 according to Table 3.1 of EN 1998-3.

Taking into account the data of Figure 7, the "*confinement effectiveness factor*" of the core $b_c \times h_c$ of the section is calculated:

$$\alpha = \left(1 - \frac{s_h}{2b_c}\right) \cdot \left(1 - \frac{s_h}{2h_c}\right) \cdot \left(1 - \frac{\sum b_i^2}{6 \cdot b_c \cdot h_c}\right) = \left(1 - \frac{6.4}{2 \cdot 34.4}\right) \cdot \left(1 - \frac{6.4}{2 \cdot 34.4}\right) \cdot \left(1 - \frac{1548}{6 \cdot 34.4 \cdot 34.4}\right) =$$

$$= 0.907 \cdot 0.907 \cdot 0.782 = 0.643$$

where

$$\sum b_i^2 = \left(9^2 + 15^2 + 9^2\right) \times 4 = 1548\,\text{cm}^2$$

The ratio of transverse steel parallel to the loading local direction y of the section is (Figure 7):

$$\rho_{sy} = \frac{A_{sw,y}}{h \cdot s_h} = \frac{4 \times \left(0.50\,\text{cm}^2\right)}{\left(45\,\text{cm} \times 6.4\,\text{cm}\right)} = 0.00694$$

The strength f_{cc} of the confined concrete and its contemporary strain ε_{cc} is given as follows:

$$f_{cc} = \frac{f_{cm}}{CF}\left[1+3.7\cdot\left(\frac{a\cdot\rho_{sy}\cdot f_{yw,m}}{f_{cm}}\right)^{0.86}\right] = \frac{43}{1.20}\left[1+3.7\cdot\left(\frac{0.643\cdot 0.00694\cdot 545}{43}\right)^{0.86}\right] =$$

$$= \frac{43}{1.20}\cdot 1.313 = 47.049\ \text{MPa}$$

$$\varepsilon_{cc} = \varepsilon_{c2}\left[1+5\cdot\left(\frac{f_{cc}}{f_{cm}/CF}-1\right)\right] = 0.002\cdot\left[1+5\cdot\left(\frac{47.049}{43/1.20}-1\right)\right] = 0.0051$$

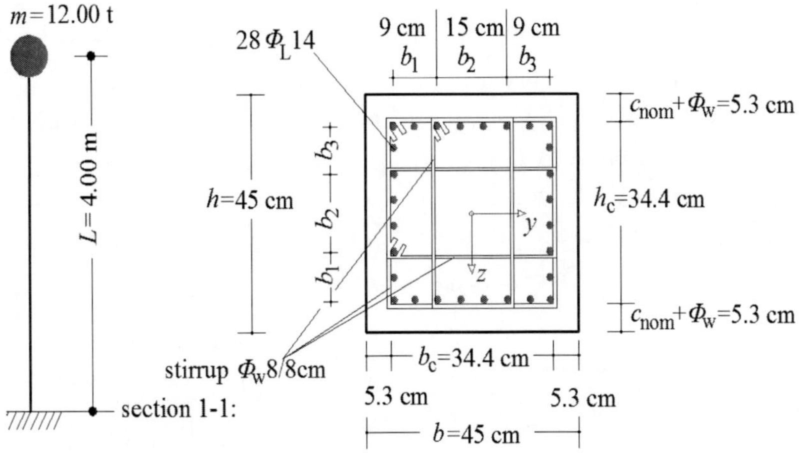

Figure 7. Reinforced concrete cantilever column with square section *h*x*b*.

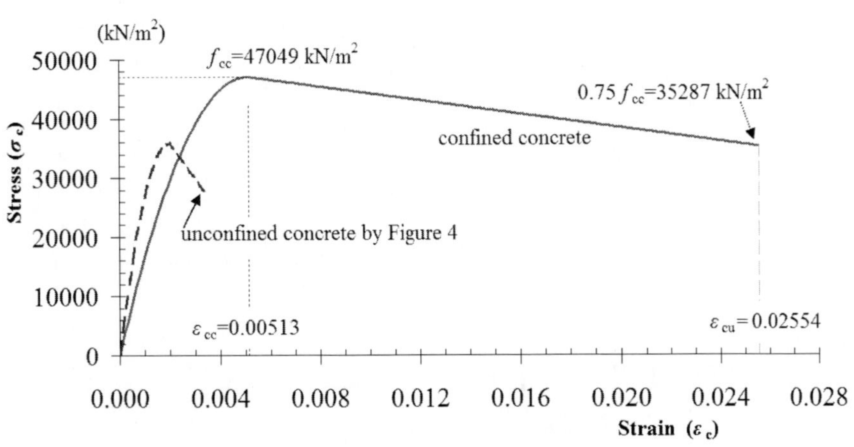

Figure 8. Stress-strain (σ - ε) diagram for confined concrete core according to EN 1998-3. Application for experimental in-situ mean strength $f_{cm} = 43\,\text{MPa}$ with Confidence Factor CF=1.20.

The ultimate strain ε_{cu} of the extreme fiber of the pressure, confined zone of the section is given:

$$\varepsilon_{cu} = 0.004 + 0.5 \cdot \frac{\alpha \cdot \rho_{sy} \cdot \left(f_{yw,m}/CF\right)}{f_{cc}} = 0.004 + 0.5 \cdot \frac{0.643 \cdot 0.00694 \cdot (545/1.20)}{47.049} = 0.02554$$

Afterwards, using Eq.(11-12) the final Stress-Strain diagram of confined concrete is calculated (Figure 8).

2.5. Available Yield Chord Rotation θ_y of a Cantilever

Afterwards of the calculation of the elastoplastic diagram of Moment-Curvature M - φ of the critical end section at the base of each cantilever, its available yield chord rotation θ_y for the "Damage Limitation" limit state can be calculated. For this purpose, the following assumption "*the behavior of the cantilever with height L_s is linear-elastic until the appearance of the yield state at its base*" is made (Figure 9). It is worthy to note that in the same Figure, the variation of the corresponding lateral yield displacement of the free-end of the cantilever, δ_y, is shown. Next, at the base of the cantilever, the yielding curvature φ_y can be calculated by $\varphi_y = M_y / EI_{eff}$ and the distribution of the curvatures follows the diagram of flexural moment in elevation, because we are located into linear and elastic area; thus at this state, the available yield chord rotation θ_y of the cantilever can be calculated elastically. For example, using the "*unit virtual load method*", the respective flexural moment at the base of the cantilever is $M_{,1} = 1 \cdot L_s$. Therefore, the yield horizontal displacement δ_y is easy calculated by the *Static of Structure*:

$$\delta_y = \frac{M_{,y} \cdot M_{,1} \cdot L_s}{3EI_{,eff}} = \frac{\varphi_y \cdot L_s^2}{3} \qquad (13)$$

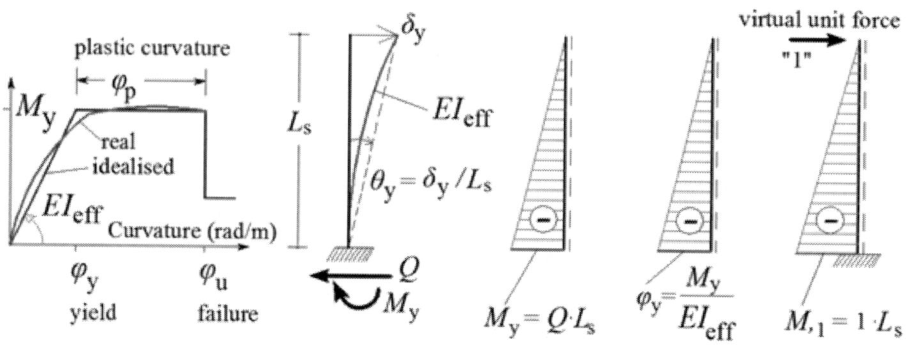

Figure 9. Linear-elastic behavior of deformed cantilever until the appearance of the yield state at its base.

Thus, the yield chord rotation of the cantilever is given:

$$\theta_y = \frac{\delta_y}{L_s} = \frac{\varphi_y \cdot L_s}{3} \tag{14}$$

However, as has been already mentioned, there are more sources that contribute in yield rotations of the end-section, such as the action of shear force and the extraction or lap-splice slip of longitudinal steel bars from the fixed-base (or the join) of the cantilever. For this reason, it is preferable to use following semi-empirical equation that is proposed by Eq.(A.10a)/sect.A.3.2.4 of Eurocode EN 1998-3, and it is based on work by Panagiotakos and Fardis (1999, 2001):

$$\theta_y = \frac{\varphi_y \cdot (L_s + a_v \cdot z)}{3} + 0.00135 \cdot \left(1 + \frac{1.50 \cdot h}{L_s}\right) + \frac{\varepsilon_y \cdot d_b \cdot (f_{ym}/CF)}{6 \cdot (d - d_1)\sqrt{f_{cm}/CF}} \tag{15}$$

where a_v is zero when the flexural failure precedes the shear failure and a_v is one when the shear failure precedes the flexural one, z is the length of internal lever arm, taken equal to $d - d_2$ in beams and columns, d and d_2 being the depths to the tension and compression reinforcement for the external compressive fiber of the section, respectively. Also, d_1 is the distance from the tension reinforcement to external tension fiber of the section, h is the depth of the geometric section of the member (Figure 2) and ε_y is the steel strain that is taken equal $\varepsilon_y = f_{ym}/(CF \cdot E_s)$.

In addition, E_s is the Elasticity Modulus of the steel, d_b is the mean diameter of steel bars of longitudinal reinforcement in meters and f_{ym} and f_{cm} are the yielding stress (mean value in MPa) of steel and concrete, respectively. Final, CF is the abovementioned Confidence Factor according to Table 3.1 of EN 1998-3, while it is worth noting that, the first part of Eq.(15) is the contribution of the flexural deformations. The second part of Eq.(15) is the contribution of shear deformations (by experimental data) and the third part is the contribution of the extraction or lap-splice slip of the longitudinal steel bars from the fixed-base of the cantilever by experimental data, too.

2.6. Available Ultimate Chord Rotation θ_u of a Cantilever

The ultimate available chord rotation θ_u of a cantilever, for the "Near Collapse" limit state under cyclic loading, can be calculated by Eq.(16) that is proposed by Eq.(A.1)/sect.A.3.2.2 of Eurocode EN 1998-3:

$$\theta_u = \frac{1}{\gamma_{el}} \cdot 0.016 \cdot 0.3^{v} \cdot \left(\frac{p_2}{p_1} \cdot \frac{f_{cm}}{CF}\right)^{0.225} \cdot a_s^{0.35} \cdot \left(25^{\alpha \cdot \rho_{sy} \cdot \frac{f_{yw,m}}{f_{cm}}}\right) \cdot \left(1.25^{100 \rho_d}\right) \tag{16}$$

where, γ_{el} is a safety factor that is taken equal to 1.50 for primary seismic structural members (due to scattering of the experimental values) and 1.00 for secondary seismic members.

$p_1 = \max(0.01,\ \omega)$ and $p_2 = \max(0.01,\ \omega')$, with ω and ω' are the mechanical reinforcement ratios of the tension reinforcement A_{s1} (with the intermediary reinforcement) and the compression one A_{s2}, respectively:

$$\omega = \rho_1 \cdot \frac{f_{ym}}{f_{cm}} = \frac{A_{s1}}{b_w \cdot d} \cdot \frac{f_{ym}}{f_{cm}} \quad , \omega' = \rho_2 \cdot \frac{f_{ym}}{f_{cm}} = \frac{A_{s2}}{b_w \cdot d} \cdot \frac{f_{ym}}{f_{cm}} \qquad (17a,b)$$

CF is the abovementioned Confidence Factor according to Table 3.1 of EN 1998-3

$\nu = \dfrac{N}{b_w \cdot h \cdot (f_{cm}/\mathrm{CF})}$ is the normalized axial force (b_w is the width of compression zone and force N is taken positive for compression, Figure 2). $\rho_d = \dfrac{A_{sd}}{b_w \cdot d}$ is the steel ratio of diagonal reinforcement A_{sd} (if exists). $a_s = M/(Q_y \cdot h) = L_s/h$ is the shear ratio at the end-section of the cantilever (Figure 2). α is the "*confinement effectiveness factor*" of the core-concrete that is given by Eq.(10):

For the case of r/c walls, the ultimate chord rotation at the limit state of "*Near Collapse*" given by Eq.(16) is divided by a factor 1.60. Moreover, the available plastic rotation θ_p is always given by $\theta_p = \theta_u - \theta_y$, while the significant chord rotation of the cantilever at limit state of "*Significant Damage*" is taken equal to with the $0.75\theta_u$ according to sect.A.3.2.3 of Eurocode EN 1998-3.

2.7. Available Cyclic Shear Strength V_R of a Cantilever

The available cyclic shear strength V_R (in MN) is decreased with the incremental demand plastic rotation $\theta_{p,d}$ according to following experimental expression according to Eq.(A.12) of Eurocode EN 1998-3:

$$V_R = \frac{1}{\gamma_{el}} \cdot \left[\frac{(h-x)\lambda_1}{2L_s} + (1-0.05\lambda_2) \cdot \left[0.16\lambda_3 (1\text{-}0.16\lambda_4) A_c \sqrt{f_{cm}/\mathrm{CF}} + V_w \right] \right] \qquad (18)$$

where, γ_{el} is a safety factor that is taken equal to 1.15 for primary seismic structural elements (due to scattering of the experimental values) and is taken 1.00 for secondary seismic members. x is the compression zone depth (in meters) that is known by the "fiber analysis" of the section (Figure 2),CF is the abovementioned Confidence Factor according to Table 3.1 of EN 1998-3.

$\lambda_1 = \min(N,\ 0.55A_c f_{cm}/CF)$, N is the axial force in MN that is positive for compression, while when the axial force is tensional then it is taken zero, $A_c = b_w d$ for rectangular sections with b_w as width of compression zone and d is the depth of the tension reinforcement in meters, f_{cm} is the concrete compressive strength (mean value) in MPa. $\lambda_2 = \min\left(5,\ \mu_\Delta^p\right)$, where $\mu_\Delta^p = \theta_p/\theta_y$. $\lambda_3 = \max(0.5,\ 100\rho_{tot})$, where ρ_{tot} is the total longitudinal reinforcement ratio (tensional, compression and intermediate), namely $\rho_{tot} = (A_{s1} + A_{s2} + A_{sv})/(b_w d)$. $\lambda_4 = \min(5,\ a_s)$, where $a_s = M/(Q_y \cdot h) = L_s/h$ with Q_y is the contemporary shear force (Figure 2). V_w is the contribution of the transverse reinforcement to shear strength, taken as being equal to $V_w = \rho_w \cdot b_w \cdot z \cdot f_{yw,m}/CF$ for cross-section with rectangular web of width b_w. ρ_w is the transverse reinforcement ratio that is given by $\rho_w = (A_{sw} \cdot \ell_w)/(h_c \cdot b_c \cdot s)$, where ℓ_w is the total length of the stirrups, A_{sw} is the steel section area of the stirrup, h_c and b_c the dimensions of the confined core of the section and s is the centerline spacing of stirrups.

2.8. Moment-Chord Rotation Diagram M-θ of the Cantilever

In order to define the final elastic-plastic diagram of Moment-Chord Rotation (M-θ) of a cantilever, it must be checked which type of failure precedes; flexure or shear? Thus, since the shear strength V_R is known by Eq.(16), the moment $M_{u,v}$ at the base of the cantilever due to V_R is easily calculated as $M_{u,v} = L_s \cdot V_R$.

When $M_{u,v}$ is greater than the flexural yielding moment M_y, then the flexural failure of the cantilever precedes the shear one. In that case, the final elastic-plastic diagram of Moment-Chord Rotation (M-θ) of a cantilever is given by Figure (10a).

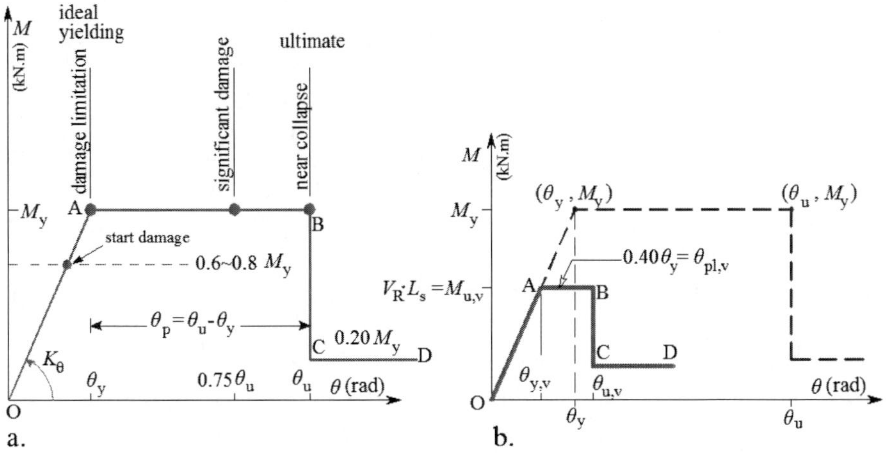

Figure 10. (a) Moment-Chord Rotation Diagram for (a) ductile failure and (b) for shear failure.

However, on the contrary, when $M_{u,v}$ is smaller than the flexural yielding moment M_y, then the shear failure of the cantilever precedes the flexural one. In the latter case, the final elastic-plastic diagram of Moment-Chord Rotation (M-θ) diagram of the cantilever is given as the curve OABCD of Figure (10b) according to sect.7.2.4.2 of KANEPE 2012.

2.9. Effective Flexural Stiffness of Member Sections

As it is clear, the abovementioned cantilever (with constant geometric dimensions along its length) has linear-elastic behavior until of the critical section at its base reaches the yielding state. Therefore, it can be concluded that the flexural stiffness $E_c I$ of the member section can be constant for the total length of the member and thus its effective value ($E_c I_{eff}$) can be calculated from Eq.(14) having known the corrected value of yield rotation θ_y by Eq.(15). Thus, the effective flexural stiffness $E_c I_{eff}$ is given according to sect.A.3.2.4(5)/ EN 1998-3:

$$E_c I_{eff} = \frac{M_y \cdot L_s}{3 \cdot \theta_y} \tag{19}$$

Therefore, in the case of a real structural member (column or beam) that has plastic hinges at its two ends, the mean effective flexural stiffness $E_c I_{eff}$ of the member-section can be estimated as the arithmetic mean of four different bend states, at the two ends of the element, for positive and negative sign of moments. This effective flexural stiffness $E_c I_{eff}$ of the member cross-section is suitable for modelling in the case of dynamic cyclic behavior, when the building is subjected to earthquake loading. It should be noted that, the abovementioned assumption about the $E_c I_{eff}$ is rational in the case where two plastic hinges are presented simultaneously at the two ends of a structural member. However, when none (or one only) plastic hinge appears on the structural member then the previous assumption is not justifiable.

When the effective flexural stiffness $E_c I_{eff}$ by Eq.(19) is taken into account for all structural members of the building model, then it is expected that the eigen-periods of vibration of the multi-degree model are changed and became longer. On the one hand, it is well-known that using this modelling there may be some mismatch at the beginning of the analysis compared to experimental results, but there is a very good agreement (with reference to seismic demand displacements and deformations) after the elements reach there damaged state.

Besides, the total procedure is Displacement (and Deformation) Based Method. On the other hand, a possible result of this alteration of the periods of the models is that, the structure's model does not load seismically adequately, because the state of resonance, between the building model and the seismic excitation is removed, since the model has high flexibility.

2.10. Numerical Application

For the abovementioned example that has been initially presented in above paragraph 2.4.1, and in order to calculate the yield chord rotation of this cantilever, the calculation of Moment-Curvature diagram has to be calculated previously. Therefore, using the fiber elements technique via software XTRACT (2007) and inserting the abovementioned Stress-Strain diagrams that are shown at Figures 4, 6 and 8, we get the Moment-Curvature diagram for compression axial force $N = m \cdot g = (12\,\text{t}) \cdot (9.81\,\text{m/s}^2) = 117.72\,\text{kN}$ (Figure 11), (with the neutral axis of section in ultimate state is located at distance x=10.32cm from the external compression section fiber).

Afterwards, using the abovementioned data, the available yield chord rotation θ_y is calculated:

$$\theta_y = \frac{\varphi_y \cdot (L_s + a_v \cdot z)}{3} + 0.00135 \cdot \left(1 + \frac{1.50 \cdot h}{L_s}\right) + \frac{\varepsilon_y \cdot d_b \cdot (f_{ym}/\text{CF})}{6 \cdot (d - d_1) \cdot \sqrt{f_{cm}/\text{CF}}} =$$

$$= \frac{0.0112 \cdot (4 + 0 \cdot 0.33)}{3} + 0.00135 \cdot \left(1 + \frac{1.50 \cdot 0.45}{4}\right) + \frac{0.00227 \cdot 0.014 \cdot (545/1.20)}{6 \cdot (0.39 - 0.06) \cdot \sqrt{43/1.20}} =$$

$$\theta_y = 0.01493 + 0.00158 + 0.00122 = 0.01773\,\text{rad}$$

Following my previous point is the calculation of the available failure chord rotation θ_u of the cantilever by Eq. (16):

$$\theta_u = \frac{1}{\gamma_{el}} \cdot 0.016 \cdot 0.3^{v} \cdot \left(\frac{p_2}{p_1} \cdot \frac{f_{cm}}{\text{CF}}\right)^{0.225} \cdot a_s^{0.35} \cdot \left(25^{\alpha \cdot \rho_{sy} \cdot \frac{f_{yw,m}}{f_{cm}}}\right) \cdot \left(1.25^{100 \rho_d}\right) =$$

$$= \frac{1}{1.50} \cdot 0.016 \cdot 0.3^{0.016} \cdot \left(\frac{0.10009}{0.16674} \cdot \frac{43}{1.20}\right)^{0.225} \cdot 8.89^{0.35} \cdot \left(25^{0.643 \cdot 0.00694 \cdot \frac{545}{43}}\right) \cdot \left(1.25^{100 \cdot 0}\right) \Rightarrow$$

$$\theta_u = \frac{1}{1.50} \cdot 0.016 \cdot 0.98092 \cdot 1.99454 \cdot 2.1484 \cdot 1.19968 \cdot 1 = 0.12102\,\text{rad}$$

where,

$$p_1 = \max(0.01,\ \omega) = \max(0.01,\ 0.16674) = 0.16674$$

$$p_2 = \max(0.01,\ \omega') = \max(0.01,\ 0.10009) = 0.10009\,,$$

because A_{s1} is the fifteen (15) steel bars, since those are located into tension area (see that neutral axis of section is 10.32 cm, Figure 11), A_{s2} is the remain nine (9) steel bars,

$$\omega = \rho_1 \cdot \frac{f_{ym}}{f_{cm}} = \frac{A_{s1}}{b_w \cdot d} \cdot \frac{f_{ym}}{f_{cm}} = \frac{15 \cdot (1.54\,\text{cm}^2)}{45 \cdot 39} \cdot \frac{545}{43} = 0.01316 \cdot 12.67 = 0.16674$$

$$\omega' = \rho_2 \cdot \frac{f_{ym}}{f_{cm}} = \frac{A_{s2}}{b_w \cdot d} \cdot \frac{f_{ym}}{f_{cm}} = \frac{9 \cdot (1.54\,\text{cm}^2)}{45 \cdot 39} \cdot \frac{545}{43} = 0.0079 \cdot 12.67 = 0.10009$$

$$\nu = \frac{N}{b_w \cdot h \cdot (f_{cm}/\text{CF})} = \frac{117.72}{0.45 \cdot 0.45 \cdot (43000/1.20)} = 0.016$$

$\rho_d = \dfrac{A_{sd}}{b_w \cdot d} = 0$, because there are not diagonal reinforcement A_{sd}

Shear ratio $a_s = M/(Q_y \cdot h) = L_s/h = 4.00/0.45 = 8.89$

Afterwards, the available plastic chord rotation θ_p is calculated:

$$\theta_p = \theta_u - \theta_y = 0.12102 - 0.01773 = 0.10329\ \text{rad}$$

The available cyclic shear strength V_R of the cantilever is calculated by Eq.(18)Q

$$V_R = \frac{1}{\gamma_{el}} \cdot \left[\frac{(h-x)\lambda_1}{2L_s} + (1-0.05\lambda_2) \cdot \left[0.16\lambda_3 (1-0.16\lambda_4) A_c \sqrt{f_{cm}/\text{CF}} + V_w \right] \right] =$$

$$= \frac{1}{1.15} \cdot \left[\frac{(0.45-0.1032) \cdot 0.11772}{2 \cdot 4.00} + (1-0.05 \cdot 5) \cdot \left[0.16 \cdot 2.106 \cdot (1-0.16 \cdot 5) \cdot 0.2025 \cdot \sqrt{43/1.20} + 1.00289 \right] \right] =$$

$$= \frac{1}{1.15} \cdot \left[0.00510 + 0.75 \cdot (0.08169 + 1.00289) \right] = 0.71176\,\text{MN} \quad (= 711.76\,\text{kN})$$

since, safety factor; $\gamma_{el} = 1.15$, height of compression zone; x=0.1032 m, total stirrup length; $\ell_w = n \cdot \left[h - 2 \cdot \left(c_{nom} + \dfrac{d_w}{2} \right) \right] = 8 \cdot \left[45 - 2 \cdot \left(4.5 + \dfrac{0.8}{2} \right) \right] = 281.6\,\text{cm}$. c_{nom} = clear (nominal) cover by unconfined concrete of external stirrup equal with 4.5cm according to EN 1992-1-1; d_w=stirrup diameter; n=number of stirrup legs (Figure 7); section area; $A_c = 0.45 \times 0.45 = 0.2025\,\text{m}^2$; plastic part of ductility factor of chord rotations (Figure 12);

$\mu_{\Delta}^{p}=\delta_{p}/\delta_{y}=\theta_{p}/\theta_{y}=0.10329/0.01773=5.82$; total longitudinal reinforcement ratio; $\rho_{tot}=A_{s,tot}/(b_{w}d)=24\cdot(1.54\,cm^{2})/(45\times 39)=0.02106$; transverse reinforcement ratio; $\rho_{w}=(A_{sw}\cdot\ell_{w})/(h_{c}\cdot b_{c}\cdot s)=(0.50\cdot 281.6)/(34.4\cdot 34.4\cdot 8)=0.01487$; parameter λ_1 dependent by axial force; $\lambda_1=\min(N,\ 0.55A_c f_{cm}/CF)=\min(0.11772,\ 0.55\cdot 0.45\cdot 0.39\cdot 43/1.20)=\min(0.11772, 3.4588)=0.11772$ parameter λ_2 dependent by μ_{Δ}^{p} ; $\lambda_2=\min(5,\ \mu_{\Delta}^{p})=\min(5,\ 5.82)=5$; parameter λ_3 dependent by ρ_{tot}; $\lambda_3=\max(0.5,\ 100\rho_{tot})=\max(0.5,\ 100\cdot 0.02106)=2.106$, parameter λ_4 dependent by shear ratio a_s; $\lambda_4=\min(5,\ a_s)=\min(5,\ 8.89)=5$

$$V_{w}=\rho_{w}\cdot b_{w}\cdot z\cdot\frac{f_{yw,m}}{CF}=0.01487\cdot 0.45\cdot(0.39-0.06)\cdot\frac{545}{1.20}=1.00289\ MN$$

Afterwards, of abovementioned data, the moment $M_{u,v}$ that causes shear failure on the cantilever is given;

$$M_{u,v}=L_{s}\cdot V_{R}=4.00\cdot 711.76=2847.04\,kNm>M_{y}=324.20\,kNm$$

(by section analysis, Figure 11).

Thus, now the final conclusion has been resulted: " *the flexural failure precedes the shear one.*"

Thus, the available capacity diagram of cantilever chord rotations is shown in Figures 12-13.

It is worthy to note that the slope $k_{\theta}=M_{y}/\theta_{y}$ of the first branch of Moment-Chord Rotation diagram (Figure 13) can be used at a spring that is inserted in the model of the structure for the exact calculation of demand rotations.

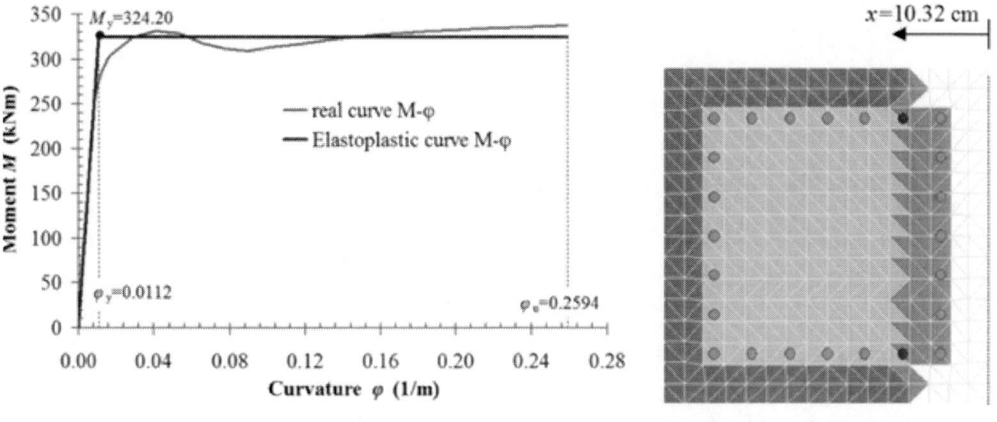

Figure 11. Section analysis by XTRACT (2007) software.

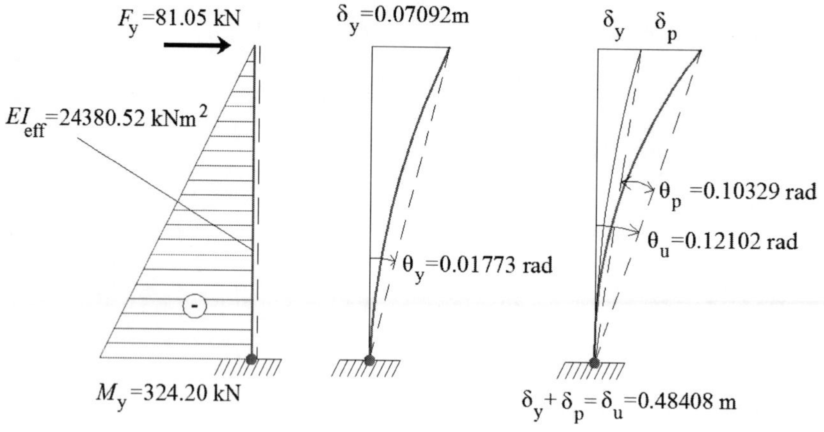

Figure 12. Cantilever profile of available displacements.

The effective flexural stiffness $E_c I_{eff}$ of the section is given:

$$E_c I_{eff} = \frac{M_y \cdot L_s}{3 \cdot \theta_y} = \frac{324.20 \cdot 4.00}{3 \cdot 0.01773} = 24380.52 \text{ kN m}^2 \quad (= 0.224 E_c I_g)$$

where I_g is the moment of inertia of geometric section.

Therefore, the effective lateral stiffness k_{eff} of the cantilever is:

$$k_{eff} = \frac{3 E_c I_{eff}}{L_s^3} = \frac{3 \cdot 24380.52}{4^3} = 1142.84 \text{ kN/m}$$

The respective yield lateral force F_y that causes yield at the cantilever base is calculated (second order effects are ignored):

$$F_y = \frac{M_y}{L_s} = \frac{324.20}{4} = 81.05 \text{ kN}$$

The yield displacement δ_y of the cantilever is:

$$\delta_y = \theta_y \cdot L_s = 0.01773 \cdot 4 = 0.07092 \text{ m}$$

or

$$\delta_y = F_y / k_{eff} = 81.05/1142.84 = 0.07092 \text{ m}$$

The failure displacement δ_u of the cantilever is:

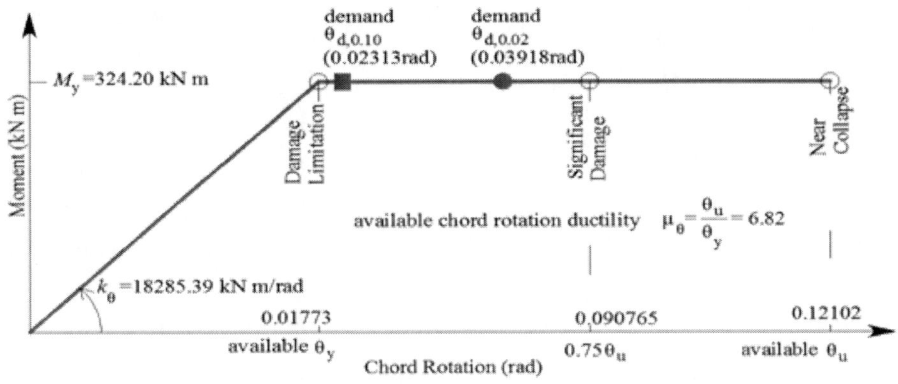

Figure 13. Available Moment-Chord Rotation diagram of cantilever.

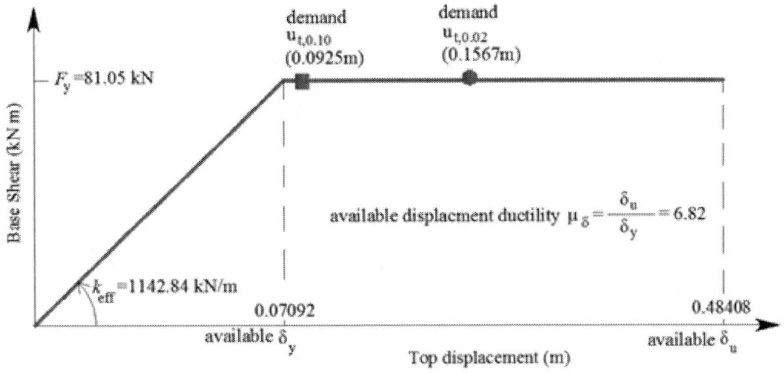

Figure 14. Available capacity curve of cantilever.

$$\delta_u = \theta_u \cdot L_s = 0.12102 \cdot 4 = 0.48408\,\text{m}$$

Thus, the available bilinear capacity curve of "base shear (F) and top displacement (δ)" of the cantilever is shown in Figure 14.

The effective period T_{eff} of the cantilever for vibration on the first branch of $F-\delta$ bilinear curve is:

$$T_{eff} = 2\pi\sqrt{\frac{m}{k_{eff}}} = 2\pi\sqrt{\frac{12}{1142.84}} = 0.644\,\text{s}$$

Last but not least is the issue of the calculation of seismic demands of the structure. For this purpose, we need various levels of design earthquakes (i.e. Design Basis Earthquake, Maximum Capable Earthquake, etc) for the specific area where the structure is founded. For example and for the needs of the present Chapter, we consider the seismic area of Greece where now we have known all necessary and contemporary seismic data (Makarios 2012b, 2012c). Moreover, two seismic scenarios will examined; (a) Design Earthquake with probability of exceedance of 2% in structure lifetime of 50 years and (b) Design Earthquake with probability of exceedance of 10% in structure lifetime of 50 years.

Table 1. New relationships between the Peak Ground Acceleration and the mean return period, for soil category *A* according to EN/1998, for the three seismic hazard zones of Greece

Greek Hazard Zone	1st relationship	2nd relationship
(low) I	$PGA_h = \dfrac{e^{(3.25+0.60 \cdot \log T_m)}}{981}$	$PGA_h = \dfrac{10^{(1.51+0.23 \cdot \log T_m)}}{981}$
(medium) II	$PGA_h = \dfrac{e^{(3.72+0.55 \cdot \log T_m)}}{981}$	$PGA_h = \dfrac{10^{(1.68+0.22 \cdot \log T_m)}}{981}$
(high) III	$PGA_h = \dfrac{e^{(4.10+0.50 \cdot \log T_m)}}{981}$	$PGA_h = \dfrac{10^{(1.82+0.20 \cdot \log T_m)}}{981}$

According to recently investigation, the new relationships between the Peak Ground Acceleration PGA_h in units of acceleration gravity g along the horizontal direction and the Mean Return Period T_m (in years) of the earthquake, for ground type A (stone) of Eurocode EN 1998-1 and for the three seismic hazard zones of Greece are given in Table 1 (Makarios, 2012c).

In this Table, two equivalent relationships are given for each Hellenic seismic hazard zone; for more accuracy reasons, we can calculate the numerical mean of the two relationships.

Moreover, the effective spectral amplification factor $\beta_{oh,eff}$ for horizontal seismic component (with damping ratio $\zeta = 0.05$) for the two abovementioned Design Earthquakes for the Hellenic conditions are given (Makarios, 2012b):

(a) Design Earthquake with probability of exceedance of 2% in structure lifetime of 50 years, $\beta_{oh,eff} = 4.10$

(b) Design Earthquake with probability of exceedance of 10% in structure lifetime of 50 years, $\beta_{oh,eff} = 3.63$

Taking into account all the abovementioned information, we get for the (a) earthquake:

$$T_m = -\frac{t}{\ln(1-P_t)} = -\frac{50}{\ln(1-0.02)} = 2475 \text{ years}$$

$$PGA_h = \frac{e^{(3.72+0.55 \cdot \log T_m)}}{981} = \frac{e^{(3.72+0.55 \cdot \log 2475)}}{981} = \frac{e^{5.5865}}{981} = 0.272g \text{ (for medium zone)}$$

Taking soil factor S=1.35 for ground type D according to Eurocode EN 1998-1, the design ground acceleration is calculated:

$$\left(PGA_{\text{h.}}\right)\cdot S = (0.272g)\cdot 1.35 = 0.3672g$$

Since $T_{\text{eff}} = 0.644\,\text{s} < T_{\text{C}} = 0.80\,\text{s}$, ($T_{\text{C}}$ is the characteristic period where the plateau of the design spectrum by EN 1998-1 for ground type D is terminated) the elastic response spectral acceleration $S_{\text{a,e}}$ for the infinitely elastic Single Degree of Freedom oscillator is calculated according to EN 1998-1:

$$S_{\text{a,e}} = \left(PGA_{\text{h}} \cdot S\right)\cdot \eta \cdot \beta_{\text{oh,eff}} = (0.3672g)\cdot 1\cdot 4.10 = 1.5055g$$

where the damping factor is $\eta = \sqrt{10/(5+\zeta)} = 1$, with equivalent viscous damping ratio $\zeta = 5\ (\%)$.

Therefore, the seismic demand elastic spectral displacement $S_{\text{d,e}}$ of the infinitely elastic single degree of freedom oscillator is given:

$$S_{\text{d,e}} = \frac{S_{\text{a,e}}}{\omega_{\text{eff}}^2} = \frac{1.5055g}{9.7565^2} = \frac{1.5055\cdot 9.81}{9.7565^2} = 0.1551\ \text{m}$$

where the circular frequency of the SDoF is $\omega_{\text{eff}} = 2\pi/T_{\text{eff}} = 2\pi/0.644 = 9.7565\ \text{rad/s}$

The seismic demand inertial force F_{el} that developed on the mass of the elastic SDoF oscillator is calculated:

$$F_{\text{el}} = m\cdot S_{\text{a,e}} = 12\cdot(1.5055\cdot 9.81) = 177.23\ \text{kN}\ >\ F_{\text{y}} = 81.05\ \text{kN}$$

Thus, the structure is going to insert in the nonlinear area for the examined design earthquake. Therefore, the reduction factor R_{y} of seismic forces is estimated:

$$R_{\text{y}} = \frac{F_{\text{el}}}{F_{\text{y}}} = \frac{177.23}{81.05} = 2.19$$

The demand displacement ductility $\mu_{\delta,0.02}$ is calculated according to unified criterion of "*equal energies*" and "*equal displacements*", which gave by Krawinkler and Nassar (1992):

$$\mu_{\delta,0.02} = 1 + \frac{R_{\text{y}}^{c} - 1}{c} = 1 + \frac{2.19^{1.044} - 1}{1.044} = 2.213$$

where the factor c is calculated:

$$c = \frac{T_{\text{eff}}}{1+T_{\text{eff}}} + \frac{0.42}{T_{\text{eff}}} = \frac{0.644}{1+0.644} + \frac{0.42}{0.644} = 1.044$$

for 0% slope of the second part of the capacity curve (elastoplastic curve) of the structure (Figure 14).

Thus, the seismic demand inelastic horizontal displacement (target-displacement) $u_{\text{t,0.02}}$ of the nonlinear SDoF oscillator is directly obtained (this procedure is known as "calculation of seismic target-displacement via nonlinear SDoF system in combine with inelastic spectra"):

$$u_{\text{t,0.02}} = \frac{\mu_{\delta,0.02}}{R_{\text{y}}} \cdot S_{\text{d,e}} = \frac{2.213}{2.19} \cdot 0.1551 = 0.1567 \text{ m}$$

The just above seismic, demand inelastic displacement $u_{\text{t,0.02}}$ gives the following seismic demand chord rotation $\theta_{\text{d,0.02}}$ of the cantilever:

$$\theta_{\text{d,0.02}} = \frac{u_{\text{t,0.02}}}{L} = \frac{0.1567}{4} = 0.03918 \text{ rad}$$

On the Figure 13, we can see that seismic demand $\theta_{\text{d,0.02}}$ is smaller than the chord rotation $0.75\theta_{\text{u}}$ of the level of significant damage, but is greater than the yield chord rotation θ_{y}, point where indicates damage.

With reference to (b) Design Earthquake with Mean Return Period $T_{\text{m}} = 475$ years, using effective spectral amplification factor $\beta_{\text{oh,eff}} = 3.63$ and repeating the previous numerical procedure, the seismic demand inelastic horizontal displacement is $u_{\text{t,0.10}} = 0.0925$ m and the seismic demand chord rotation is $\theta_{\text{d,0.10}} = 0.02313$ rad, respectively. In other words, for the (b) Design Earthquake, the structure has been just yielded, practically, as shown on Figure 14.

It is worthy to note that, in planar frames or spatial buildings, which are more complex structures than a cantilever, each plastic hinge (or each sub-cantilever I or II of Figure 1a) is checked in terms of chord rotations (Figure 13) and more details there are in below paragraph 3 of the present Chapter.

On the contrary, the available capacity curve of the structure is taken place once time, for whole structure, in terms "base shear versus horizontal displacement at the top of the structure" according to equivalent optimum Nonlinear Single Degree of Freedom system that is used in each time (Makarios, 2009, 2010, 2012a).

2.11. Modelling of Masonry Infill Walls

According to the guidelines of the KANEPE (2012), the modelling technique for masonry infill walls that will be adopted depends on the selection of the seismic performance level for which the structure will be checked. In particular:

- For the "*Damage Limitation*" limit state: In this case, the behavior of the structure is considered practically linear-elastic, thus, the masonry infill walls can be modeled with two equivalent diagonal bars, with simple hinges at their ends and with linear behavior. According to the specifications of KANEPE 2012, each bar must have rectangular cross-section and axial stiffness $EA_{\text{eff}} = 0.50 E_{\text{w}} A_{\text{w}}$, where $A_{\text{w}} = w \cdot t_{\text{w}}$ is the section area of the equivalent bar, w is the effective width (see below), t_{w} is the effective thickness of the wall and E_{w} is the Elasticity Modulus of the masonry infill wall.
- For "*Significant Damage*" limit state: In this case, the behavior of the structure is non-linear, thus, the masonry infill walls can be modeled by two equivalent diagonal bars with simple hinges at their ends and with one-sided (in compression only) nonlinear behavior. According to the specifications of KANEPE 2012, each diagonal bar must have a rectangular cross-section with axial-stiffness (in compression only) equal to $EA_{\text{eff}} = 0.68 E_{\text{w}} A_{\text{w}}$.
- For "*Near Collapse*" limit state, according to sect.7.4.1b of KANEPE 2012, all masonry infill walls should be ignored from the structural model, but on the other hand, it has minor sense because the main target is always the level of "*Significant Damage*". According to our opinion, the masonry infill walls should be taking into account in the model for the "*Near Collapse*" limit state as it happens for "*Significant Damage*" one.

According to international literature, the effective width w of the equivalent diagonal bar of a masonry infill wall can be calculated using the following equation:

$$w = (0.5) \cdot 0.175 \cdot L \cdot \lambda \cdot \lambda_{\text{h}}^{-0.4} \tag{20}$$

where, L is the length of the diagonal direction of each masonry infill wall and λ_{h} is a factor that can be calculated by Eq.(21) (Mainstone and Weeks, 1970; Mainstone, 1974; FEMA-274; FEMA-306), while λ is a reduction factor that depends on size of the opening that may exists, while it is given by Eq.(22) (Asteris, 2003, 2005 and 2008; Asteris *et al* 2011a, 2011b, 2011c, 2012).

$$\lambda_{\text{h}} = h \cdot \sqrt[4]{\frac{E_{\text{w}} \cdot t_{\text{w}} \cdot \sin 2\alpha}{4 E_{\text{c}} I_{\text{eff,m}} \cdot h_{\text{w}}}} \tag{21}$$

where $E_c I_{eff,m}$ is the arithmetic mean of the effective flexural stiffness of the two column sections that are given by Eq.(19), h is the storey high, h_w is the masonry infill wall pure high, α is the slope (referring to horizontal direction) of the diagonal bar.

$$\lambda = 1 - 2\alpha_w^{0.54} + \alpha_w^{1.14} \tag{22}$$

where $\alpha_w = A_{open} / A_{wall}$, A_{open} is the area of opening and A_{wall} is the masonry infill wall area.

The factor (0.5) of Eq.(20) is taking according to secti. 4.3.1(7) of Eurocode EN 1998-1. Moreover, it is known that the stability of an infill wall is reduced due to its vibration along axis perpendicular to its plane, (phenomenon of infill wall vibration out of its plane). Taking into account the last phenomenon and according to our opinion, all infill masonry axial stiffness should be taken to be equal to one-half of their estimated results and it is covered from the factor (0.5) of Eq.(20) for each one seismic performance level.

It is worth noting that, according to the guidelines of KANEPE 2012, the Modulus of Elasticity E_w of the masonry infill wall can be estimating by Eq.(23):

$$E_w = K_E \cdot f_{wc} \tag{23}$$

where, K_E is a factor between 500 and 1000, f_{wc} is the mean compressive strength of the masonry infill wall (in MPa), along the diagonal direction. Approximately, from the combination of EN 1998-3 and KANEPE 2012, the value of f_{wc} can be estimated via the following relationship:

$$f_{wc} = \lambda_s \cdot \lambda_m \cdot \lambda_c \cdot k \cdot f_{bc}^{0.7} \cdot f_{mc}^{0.3} / \mathrm{CF} \tag{24}$$

where, λ_s is a factor that is taken equal to 0.7 and via this factor the masonry infill wall's lateral force is converted to diagonal force of the wall, λ_m is a factor that is taken equal to 1.5 and via this factor the characteristic strength of the masonry infill wall is converted to mean strength, λ_c is a factor that is taken equal to 1.2 and via this factor the wall's strength is increased thanks to bounding r/c frame consisting of the two columns and a beam. k is a factor with value between 0.35 and 0.55 and is dependent on the bricks and mortar, f_{bc} is the mean compressive strength of the brick (about 5.5MPa for a common Greek brick), f_{mc} is the mean compressive strength of the mortar (about 3.5 MPa for a common Greek mortar), CF is the Confidence Factor according to Table 3.1 of EN 1998-3.

For "Significant Damage" limit state, according to KANEPE 2012, the one-sided nonlinear stress-strain diagram σ-ε of the compressive diagonal bar of the masonry infill wall can be represented by the one shown in Figure 15.

Therefore, the yielding axial force is $N_y = \varepsilon_y \cdot EA_{eff}$, while the yielding lengthening is $\Delta\ell_y = \varepsilon_y \cdot L_w$ and the failure lengthening is $\Delta\ell_u = \varepsilon_u \cdot L_w$ (Makarios and Asteris, 2012).

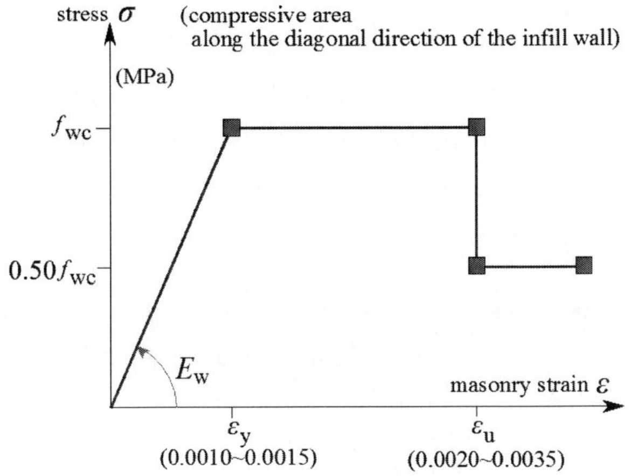

Figure 15. One-sided nonlinear diagram σ - ε of the compressive diagonal bar of the masonry infill wall.

3. SEISMIC DEMAND CHORD ROTATIONS OF STRUCTURAL MEMBERS OF BUILDING

3.1. Exterior Roof Joint in Single-Storey Frame

In the past, many works, with different accuracy in their methodologies, published with reference to calculation of seismic demand plastic rotations of r/c structural members (Kanaan and Powell, 1975; Litton, 1975; Gupta and Krawinkler, 1999; Goel and Chopra 2004; Eom *et al*, 2012). However, in the present Chapter, we are going to see the same issue from a different point of view without use inelastic springs.

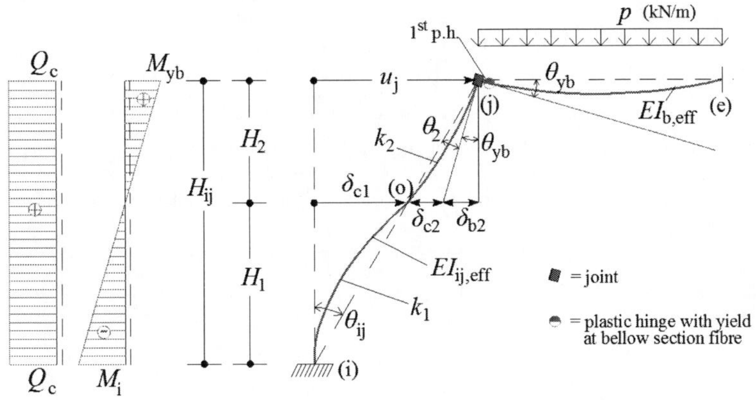

Figure 16. First plastic hinge at beam end at exterior roof joint under lateral loading.

Consider the column ij with height H_{ij} (Figure 16), which possesses section effective flexural stiffness $EI_{ij,eff}$, constant at whole its length. Moreover, the beam possesses section effective flexural stiffness $EI_{b,eff}$, where E is the Modulus of Elasticity of the concrete and $I_{ij,eff}$, $I_{b,eff}$ are the effective moment of inertia of column and beam, respectively. The section effective flexural stiffness $EI_{b,eff}$ and $EI_{ij,eff}$ are calculated by abovementioned Eq.(19) according to to sect.A.3.2.4(5)/ EN 1998-3. In addition, we consider that the beam is loaded with the permanent gravity load p, while the horizontal displacement u_j is gradually increased (pushover analysis) and points (o) and (e) are the null-moment points at column and beam, respectively.

It is clear that the position of these later two points is altered in each analysis step; moreover, the yielding flexural moment M_{yb} at the beam-end is known by the section analysis using fiber elements (i.e. see Figure 11).

3.1.1. First Plastic Hinge at Beam-End

Consider the horizontal relative displacement $u_j = u_{y1}$ for the appearance of the first plastic hinge at the beam-end (Figure 16). In that case, the flexural moment at the top of the column is M_{yb}, because the equilibrium of the joint is true, while the column diagrams of moment (M) and shear force (Q) are given in the same Figure. Afterwards, the seismic, demand column chord rotation θ_{ij} at its bottom is given:

$$\theta_{ij} = \frac{u_j}{H_{ij}} \tag{25}$$

If moment M_i at the column bottom is less than yielding column moment M_{yc}, namely $M_i < M_{yc}$, none plastic hinge appears, and, therefore, the seismic, demand column chord rotation θ_{ij} at the column bottom is less than available column chord rotation θ_{yc} that resulted by Eq.(15).

It is worth noting that in the present case, plastic hinge at the column top cannot be appeared, because there is plastic hinge at the beam-end. Thus, the seismic, demand column chord rotation θ_2 is always less than available column chord rotation θ_{yc} at the column top. According to geometry of Figure 16, the seismic, demand column chord rotation θ_2 is calculated:

$$\theta_2 = \frac{\delta_{c2}}{H_2} \tag{26}$$

Therefore, the yielding axial force is $N_y = \varepsilon_y \cdot EA_{eff}$, while the yielding lengthening is $\Delta\ell_y = \varepsilon_y \cdot L_w$ and the failure lengthening is $\Delta\ell_u = \varepsilon_u \cdot L_w$ (Makarios and Asteris, 2012).

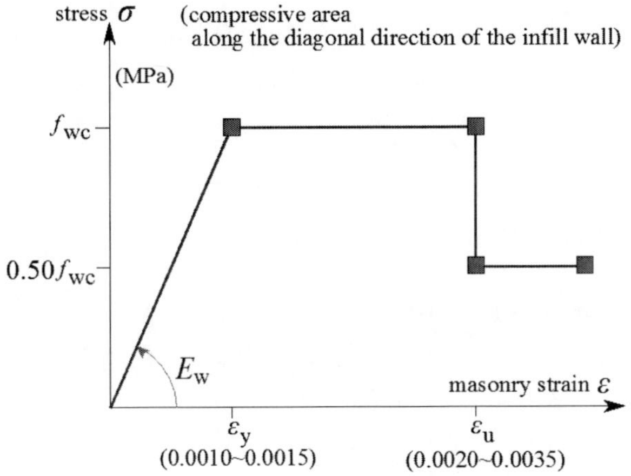

Figure 15. One-sided nonlinear diagram σ - ε of the compressive diagonal bar of the masonry infill wall.

3. Seismic Demand Chord Rotations of Structural Members of Building

3.1. Exterior Roof Joint in Single-Storey Frame

In the past, many works, with different accuracy in their methodologies, published with reference to calculation of seismic demand plastic rotations of r/c structural members (Kanaan and Powell, 1975; Litton, 1975; Gupta and Krawinkler, 1999; Goel and Chopra 2004; Eom *et al*, 2012). However, in the present Chapter, we are going to see the same issue from a different point of view without use inelastic springs.

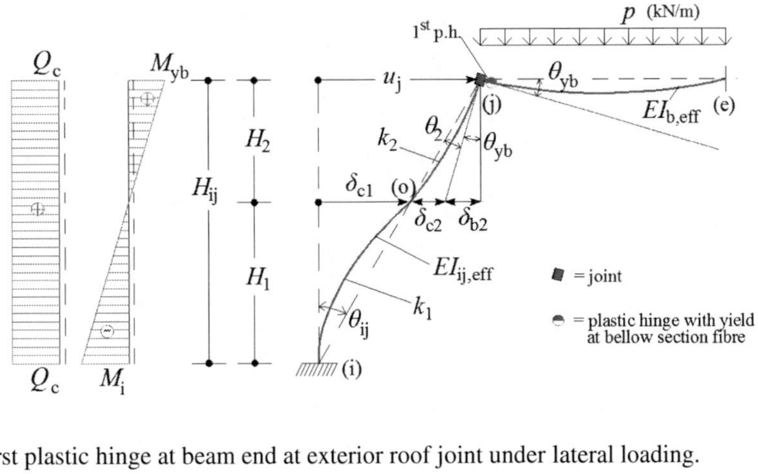

Figure 16. First plastic hinge at beam end at exterior roof joint under lateral loading.

Consider the column ij with height H_{ij} (Figure 16), which possesses section effective flexural stiffness $EI_{ij,eff}$, constant at whole its length. Moreover, the beam possesses section effective flexural stiffness $EI_{b,eff}$, where E is the Modulus of Elasticity of the concrete and $I_{ij,eff}$, $I_{b,eff}$ are the effective moment of inertia of column and beam, respectively. The section effective flexural stiffness $EI_{b,eff}$ and $EI_{ij,eff}$ are calculated by abovementioned Eq.(19) according to to sect.A.3.2.4(5)/ EN 1998-3. In addition, we consider that the beam is loaded with the permanent gravity load p, while the horizontal displacement u_j is gradually increased (pushover analysis) and points (o) and (c) are the null-moment points at column and beam, respectively.

It is clear that the position of these later two points is altered in each analysis step; moreover, the yielding flexural moment M_{yb} at the beam-end is known by the section analysis using fiber elements (i.e. see Figure 11).

3.1.1. First Plastic Hinge at Beam-End

Consider the horizontal relative displacement $u_j = u_{y1}$ for the appearance of the first plastic hinge at the beam-end (Figure 16). In that case, the flexural moment at the top of the column is M_{yb}, because the equilibrium of the joint is true, while the column diagrams of moment (M) and shear force (Q) are given in the same Figure. Afterwards, the seismic, demand column chord rotation θ_{ij} at its bottom is given:

$$\theta_{ij} = \frac{u_j}{H_{ij}} \tag{25}$$

If moment M_i at the column bottom is less than yielding column moment M_{yc}, namely $M_i < M_{yc}$, none plastic hinge appears, and, therefore, the seismic, demand column chord rotation θ_{ij} at the column bottom is less than available column chord rotation θ_{yc} that resulted by Eq.(15).

It is worth noting that in the present case, plastic hinge at the column top cannot be appeared, because there is plastic hinge at the beam-end. Thus, the seismic, demand column chord rotation θ_2 is always less than available column chord rotation θ_{yc} at the column top. According to geometry of Figure 16, the seismic, demand column chord rotation θ_2 is calculated:

$$\theta_2 = \frac{\delta_{c2}}{H_2} \tag{26}$$

Therefore, the yielding axial force is $N_y = \varepsilon_y \cdot EA_{eff}$, while the yielding lengthening is $\Delta\ell_y = \varepsilon_y \cdot L_w$ and the failure lengthening is $\Delta\ell_u = \varepsilon_u \cdot L_w$ (Makarios and Asteris, 2012).

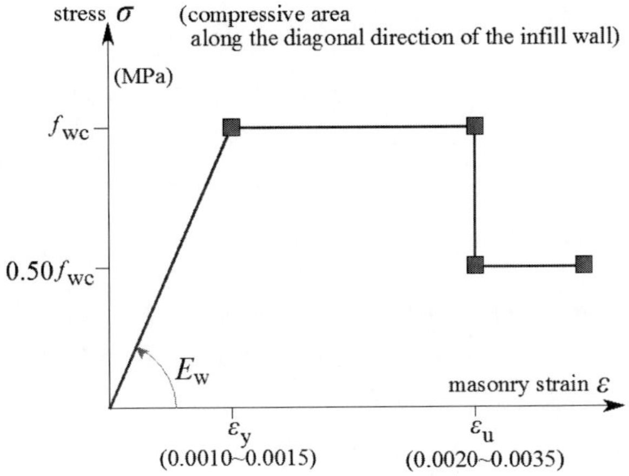

Figure 15. One-sided nonlinear diagram σ - ε of the compressive diagonal bar of the masonry infill wall.

3. Seismic Demand Chord Rotations of Structural Members of Building

3.1. Exterior Roof Joint in Single-Storey Frame

In the past, many works, with different accuracy in their methodologies, published with reference to calculation of seismic demand plastic rotations of r/c structural members (Kanaan and Powell, 1975; Litton, 1975; Gupta and Krawinkler, 1999; Goel and Chopra 2004; Eom *et al*, 2012). However, in the present Chapter, we are going to see the same issue from a different point of view without use inelastic springs.

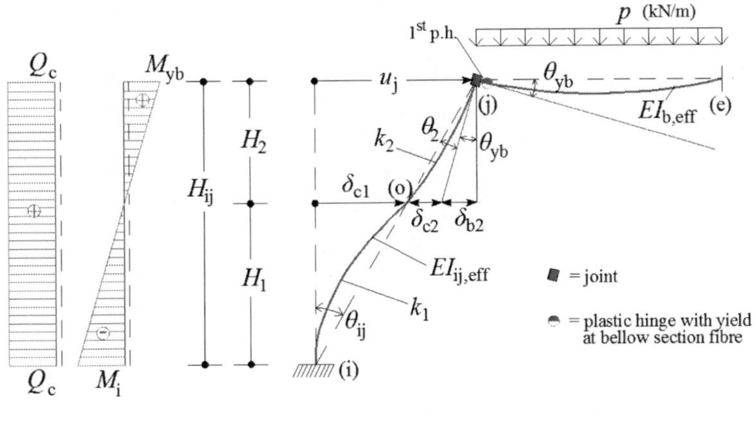

Figure 16. First plastic hinge at beam end at exterior roof joint under lateral loading.

Consider the column ij with height H_{ij} (Figure 16), which possesses section effective flexural stiffness $EI_{ij,eff}$, constant at whole its length. Moreover, the beam possesses section effective flexural stiffness $EI_{b,eff}$, where E is the Modulus of Elasticity of the concrete and $I_{ij,eff}$, $I_{b,eff}$ are the effective moment of inertia of column and beam, respectively. The section effective flexural stiffness $EI_{b,eff}$ and $EI_{ij,eff}$ are calculated by abovementioned Eq.(19) according to to sect.A.3.2.4(5)/ EN 1998-3. In addition, we consider that the beam is loaded with the permanent gravity load p, while the horizontal displacement u_j is gradually increased (pushover analysis) and points (o) and (e) are the null-moment points at column and beam, respectively.

It is clear that the position of these later two points is altered in each analysis step; moreover, the yielding flexural moment M_{yb} at the beam-end is known by the section analysis using fiber elements (i.e. see Figure 11).

3.1.1. First Plastic Hinge at Beam-End

Consider the horizontal relative displacement $u_j = u_{y1}$ for the appearance of the first plastic hinge at the beam-end (Figure 16). In that case, the flexural moment at the top of the column is M_{yb}, because the equilibrium of the joint is true, while the column diagrams of moment (M) and shear force (Q) are given in the same Figure. Afterwards, the seismic, demand column chord rotation θ_{ij} at its bottom is given:

$$\theta_{ij} = \frac{u_j}{H_{ij}} \tag{25}$$

If moment M_i at the column bottom is less than yielding column moment M_{yc}, namely $M_i < M_{yc}$, none plastic hinge appears, and, therefore, the seismic, demand column chord rotation θ_{ij} at the column bottom is less than available column chord rotation θ_{yc} that resulted by Eq.(15).

It is worth noting that in the present case, plastic hinge at the column top cannot be appeared, because there is plastic hinge at the beam-end. Thus, the seismic, demand column chord rotation θ_2 is always less than available column chord rotation θ_{yc} at the column top. According to geometry of Figure 16, the seismic, demand column chord rotation θ_2 is calculated:

$$\theta_2 = \frac{\delta_{c2}}{H_2} \tag{26}$$

where, $H_2 = M_{yb}/Q_c$, $\delta_{c2} = Q_c/k_2$ and k_2 is the lateral stiffness of the "ideal cantilever j-o" that is given:

$$k_2 = \frac{3EI_{ij,eff}}{H_2^3} \tag{27}$$

Thus, the seismic demand beam yielding rotation θ_{yb} of the beam-end is given as function of the horizontal relative yielding displacement u_j of the single-storey frame (Figure 16):

$$\theta_{yb} = \theta_{ij} - \theta_2 = \frac{u_j}{H_{ij}} - \frac{\delta_{c2}}{H_2} \tag{28}$$

3.1.2. Seismic Demand Plastic Rotation θ_{pb} at the Beam-End for $u_{dem} > u_y$

Consider an advanced analysis step, where the horizontal relative displacement u_{dem} is greater than the horizontal relative yielding displacement u_{y1} (Figure 17). In that case, the seismic, demand column chord rotation $\theta_{ij'}$ at the column bottom is:

$$\theta_{ij'} = \frac{u_{dem}}{H_{ij}} \tag{29}$$

In order to calculate the seismic demand plastic rotation θ_{pb} of the first plastic hinge at the beam-end, the following two separated cases must be examined:

- Case (i): Before the appearance of the second plastic hinge at column bottom, namely $M_i \le M_{yc}$ and $\theta_{1j'} \le \theta_{yc}$;
- Case (ii): After the appearance of the second plastic hinge at column bottom, namely, namely $M_i = M_{yc}$ and $\theta_{ij'} \ge \theta_{yc}$;

With reference to Case (i), we consider that after the appearance of the first plastic hinge at the beam-end, the $M_{i'} \le M_{yc}$ and $\theta_{ij'} \le \theta_{yc}$ is true (Figure 17). Then, the chord rotation $\theta_{2'}$ at the top of the column is calculated geometrically:

$$\theta_{2'} = \frac{\delta_{c2'}}{H_{2'}} \tag{30}$$

where, $H_{2'} = M_{yb}/Q_{c'}$, $\delta_{c2'} = Q_{c'}/k_{2'}$ and $k_{2'} = 3EI_{ij,eff} / H_{2'}^3$.

Therefore, the total rotation $\theta_{b,tot}$ of the joint (j), as function of the seismic demand horizontal relative displacement u_{dem}, is given:

$$\theta_{b,tot} = \theta_{ij'} - \theta_{2'} = \frac{u_{dem}}{H_{ij}} - \frac{\delta_{c2'}}{H_{2'}} \tag{31}$$

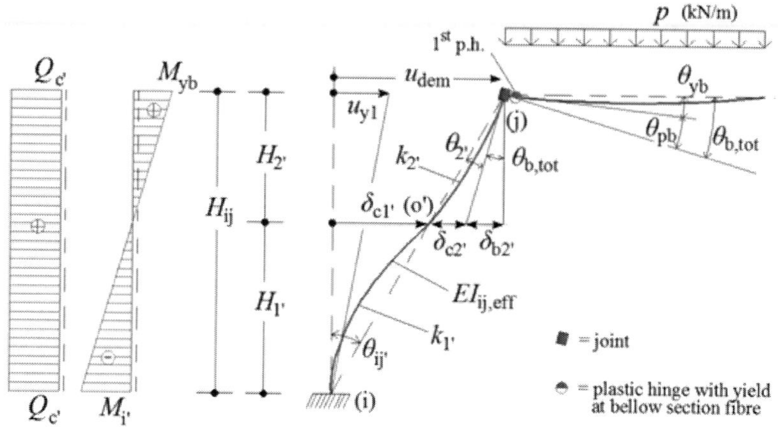

Figure 17. Seismic demand plastic beam rotation for a known demand storey drift u_{dem} before the appearance of a column plastic hinge.

Thus, the seismic demand beam plastic rotation θ_{pb} that is developed into the first plastic hinge of the beam-end is given (Figure 17):

$$\theta_{pb} = \theta_{b,tot} - \theta_{yb} = \left(\frac{u_{dem}}{H_{ij}} - \frac{\delta_{c2'}}{H_{2'}}\right) - \left(\frac{u_{y1}}{H_{ij}} - \frac{\delta_{c2}}{H_2}\right) \Rightarrow \tag{32}$$

$$\theta_{pb} = \frac{u_{dem} - u_{y1}}{H_{ij}} - \frac{\delta_{c2'}}{H_{2'}} + \frac{\delta_{c2}}{H_2} \tag{33}$$

Afterwards, the checking of seismic performance matrix (Figure 10(a)) can be performed, where according to EN 1998-3, if $\left(\theta_{yb} + \theta_{pb}\right) \le 0.75\theta_{ub}$ is true then the seismic demand is satisfied for the level "significant damage", while if $\left(\theta_{yb} + \theta_{pb}\right) < \theta_{ub}$ is true then the seismic demand is satisfied for the level "near collapse".

With reference to case (ii), we consider that after the appearance of the first plastic hinge at the beam-end, the second plastic hinge is appeared at the column bottom. For more greater lateral relative displacement u_{dem}, it is clear that $M_i = M_{yc}$ and $\theta_{ij''} > \theta_{yc}$ is true (Figure 18). In that case, the chord rotation $\theta_{2''}$ is given by $\theta_{2''} = \delta_{c2''}/H_{2''}$, where $H_{2''} = M_{yb}/Q_{c''}$, $\delta_{c2''} = Q_{c''}/k_{2''}$ and $k_{2''} = 3EI_{ij,eff}/H_{2''}^3$.

Thus, the total rotation of joint (j), as function of the seismic demand horizontal relative displacement u_{dem} is given:

$$\theta_{b,tot} = \theta_{ij"} - \theta_{2"} = \frac{u_{dem}}{H_{ij}} - \frac{\delta_{c2"}}{H_{2"}} \tag{34}$$

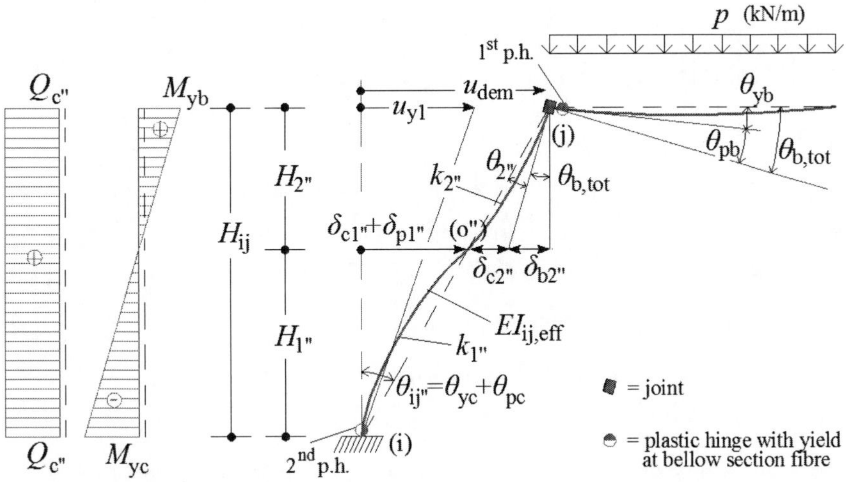

Figure 18. Seismic demand plastic beam rotation for a given demand storey drift u_{dem} after the appearance of the column plastic hinge.

Then, the demand plastic rotation θ_{pb} of the first plastic hinge at the beam-end is calculated:

$$\theta_{pb} = \theta_{b,tot} - \theta_{yb} = \left(\frac{u_{dem}}{H_{ij}} - \frac{\delta_{c2"}}{H_{2"}}\right) - \left(\frac{u_{y1}}{H_{ij}} - \frac{\delta_{c2}}{H_2}\right) \Rightarrow \tag{35}$$

$$\theta_{pb} = \frac{u_{dem} - u_{y1}}{H_{ij}} - \frac{\delta_{c2"}}{H_{2"}} + \frac{\delta_{c2}}{H_2} \tag{36}$$

Afterwards, the checking of seismic performance matrix can be performed. However, in that case, the plastic hinge at the column bottom must be checked. For this purpose, we consider the inelastic ideal cantilever i-o" (as part of the column), where the free-end o" has the elastic displacement $\delta_{c1"} = Q_{c"}/k_{1"}$ and the unknown plastic displacement $\delta_{p1"}$, with $k_{1"} = 3EI_{ij,eff} / H_{1"}^3$ is the lateral stiffness of the ideal cantilever i-o". Then, for known u_{dem} , at the column null-moment point o" , the demand plastic displacement $\delta_{p1"}$ is calculated:

$$\delta_{p1"} = u_{dem} - \delta_{c1"} - \delta_{c2"} - \delta_{b2"} \tag{37}$$

where $\delta_{b2"} = \theta_{b,tot} \cdot H_{2"}$.

Thus, the demand plastic rotation θ_{pc} of the column plastic hinge is given:

$$\theta_{pc} = \frac{\delta_{p1"}}{H_{1"}} \tag{38}$$

In addition, the checking of seismic performance matrix can be performed.

3.2. Interior Roof Joint in Single-Storey Frame

3.2.1. First Plastic Hinge at Beam-End

Consider that the yielding moments M_{yb1} and M_{yb2} at the two beam-ends (of beams I and II, respectively) that converge at joint (j) are known (Figure 19).

Moreover, we consider the horizontal relative displacement $u_j = u_{y1}$ that takes place when the first plastic hinge (with yield moment M_{yb1}) is appeared at the end of beam I. Then, the seismic demand column chord rotation θ_{ij} at the column bottom is directly given by $\theta_{ij} = u_j / H_{ij}$, which is less than available column chord rotation θ_{yc}. Then, at the top of the column, the column chord rotation θ_2 can be calculated geometrically by $\theta_2 = \delta_{c2}/H_2$, where $H_2 = M_{cj}/Q_c$, $\delta_{c2} = Q_c/k_2$ and $k_2 = 3EI_{ij,eff} / H_2^3$. Therefore, the seismic, demand beam yield rotation θ_{yb} of the beam-end is calculated as a function of the horizontal relative displacement u_j of the single-storey frame:

$$\theta_{yb} = \theta_{ij} - \theta_2 = \frac{u_j}{H_{ij}} - \frac{\delta_{c2}}{H_2} \tag{39}$$

3.2.2. Demand Plastic Rotation θ_{pb} at the Beam-End for $u_{dem} > u_{y1}$

Afterwards the first plastic hinge at the beam and for horizontal relative displacement, $u_{dem} > u_{y1}$, the plastic rotation θ_{pb} that takes place into the first plastic hinge can be calculated (Figure 20).

Indeed, in that case, the seismic demand chord rotation at the column bottom is . Moreover, now, the main question is that; how can calculate the demand beam plastic rotation for in the following cases?

- Case (i): when the second plastic hinge (at beams or column) has not been appeared.
- Case (ii): when the second plastic hinge has been appeared at the end of the beam II.
- Case (iii): when the second plastic hinge has been appeared at the column bottom.
- Case (iv): when the second plastic hinge has been appeared at the top of the column.

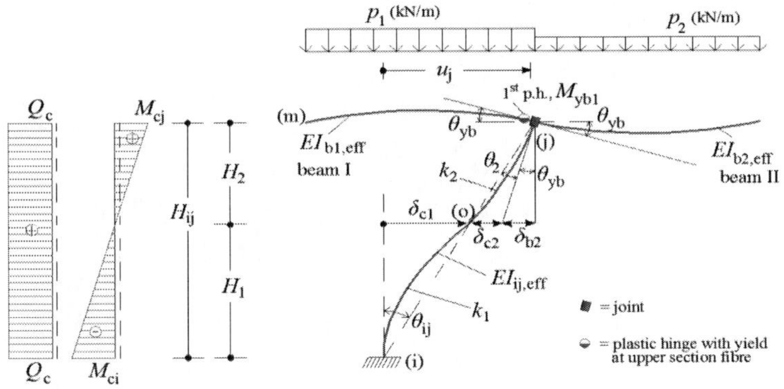

Figure 19. Appearance of the first plastic hinge at beam end at interior roof joint of a single-storey frame under lateral loading.

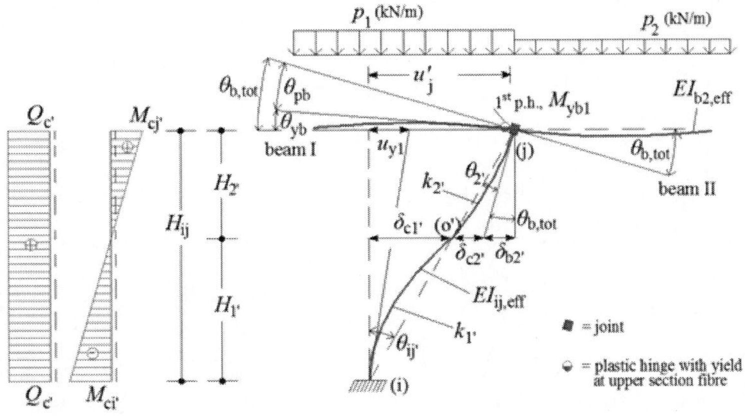

Figure 20. Demand plastic beam rotation θ_{pb} for a known demand storey drift u_{dem} before the appearance of a column plastic hinge.

If cases (i) or (ii) are validated then Eqs.(30-33) are directly used; If case (iii) is true then Eqs.(34-38) are directly used. If case (iv) is true then we get Figure 21, where when the second plastic hinge is appeared at the top of the column then the horizontal relative displacement is u_{y2}. In the last case, the seismic demand chord rotation at the column bottom is $\theta_{ij''} = u_{y2}/H_{ij}$. The seismic demand chord rotation $\theta_{2''}$ is geometrically calculated as $\theta_{2''} = \delta_{c2''}/H_{2''}$, where $H_{2''} = M_{ycj}/Q_{c''}$, $\delta_{c2''} = Q_{c''}/k_{2''}$ and $k_{2''} = 3EI_{ij,eff}/H_{2''}^3$. Then, the total rotation $\theta_{b,tot}$ of joint (j) is given:

$$\theta_{b,tot} = \theta_{ij''} - \theta_{2''} = \frac{u_{y2}}{H_{ij}} - \frac{\delta_{c2''}}{H_{2''}} \tag{40}$$

Thus, the seismic demand plastic rotation θ_{pb} of the first plastic hinge at the beam-end is given (Figure 21):

$$\theta_{pb} = \theta_{b,tot} - \theta_{yb} = \left(\frac{u_{y2}}{H_{ij}} - \frac{\delta_{c2"}}{H_{2"}}\right) - \left(\frac{u_{y1}}{H_{ij}} - \frac{\delta_{c2}}{H_2}\right) \Rightarrow \tag{41}$$

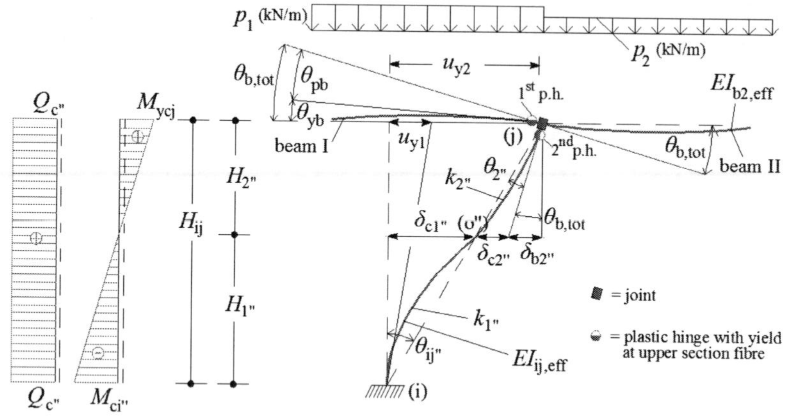

Figure 21. Demand plastic beam rotation θ_{pb} at time where the 2nd plastic hinge is appeared at the top of the column.

$$\theta_{pb} = \frac{u_{y2} - u_{y1}}{H_{ij}} - \frac{\delta_{c2"}}{H_{2"}} + \frac{\delta_{c2}}{H_2} \tag{42}$$

After that, for more greater horizontal relative displacement $u_{dem} > u_{y2}$, the column does not contribute τo joint stiffness and it means that the demand beam plastic rotation θ_{pb} remain constant, without increase, beside the fact where the horizontal relative displacement u_{dem} is more greater .

3.3. Exterior and Interior Floor Joint in Multi-Storey Frame

3.3.1. First Plastic Hinge at Beam-End

Consider that, just as the 1st plastic hinge is appeared at the beam-end of *j*-floor then the horizontal floor displacements u_i, u_j and u_k of *i*-, *j*- and *k*-floor are known, respectively (Figures 22, 23). In that case, the seismic demand chord rotations θ_{ij} and θ_{jk} of columns ij and jk, respectively, are given:

$$\theta_{ij} = \frac{u_j - u_i}{H_{ij}} \tag{43}$$

$$\theta_{jk} = \frac{u_k - u_j}{H_{jk}} \tag{44}$$

Then, at the top of column ij, its chord rotation is $\theta_2 = \delta_{c2}/H_2$, where $H_2 = M_{c2}/Q_{c2}$, $\delta_{c2} = Q_{c2}/k_2$ and $k_2 = 3EI_{ij,eff} \big/ H_2^3$, (Figures 22, 23). Moreover, at the base of column jk, its chord rotation is $\theta_3 = \delta_{c3}/H_3$, where $H_3 = M_{c3}/Q_{c3}$, $\delta_{c3} = Q_{c3}/k_3$ and $k_3 = 3EI_{jk,eff} \big/ H_3^3$. Thus, the seismic, demand yield chord rotation θ_{yb} of the beam-end is given:

$$\theta_{yb} = \theta_{ij} - \theta_2 = \theta_{jk} - \theta_3 \tag{45}$$

or more accuracy:

$$\theta_{yb} = \frac{(\theta_{ij} - \theta_2) + (\theta_{jk} - \theta_3)}{2} \tag{46}$$

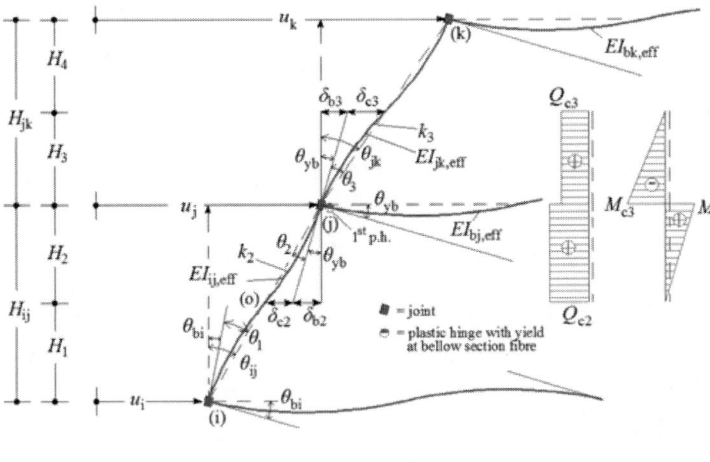

Figure 22. Appearance of the first plastic hinge at beam end at exterior floor joint.

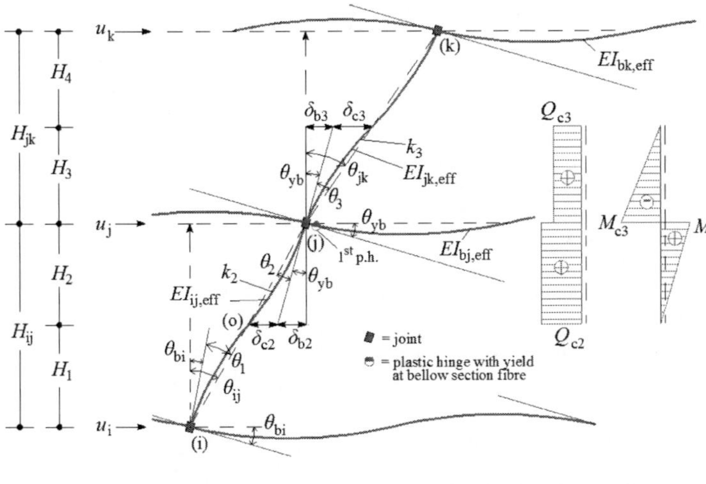

Figure 23. Appearance of the first plastic hinge at beam end at interior floor joint.

3.3.2. Demand Plastic Chord Rotation θ_{pb} at Beam-End

Afterwards the appearance of the 1st plastic hinge, we consider that the r/c frame appears more large seismic demand horizontal floor displacements u'_i, u'_j and u'_k of *i*-, *j*- and *k*-floor, respectively (Figures 24-26), where the 2nd plastic hinge is appeared somewhere, as say, at the base of column jk.

The chord rotations $\theta_{ij'}$ and $\theta_{jk'}$ of columns ij and jk, respectively, are given:

$$\theta_{ij'} = \frac{u'_j - u'_i}{H_{ij}} \tag{47}$$

$$\theta_{jk'} = \frac{u'_k - u'_j}{H_{jk}} \tag{48}$$

Then, at the bottom of the column ij, its chord rotation is $\theta_{2'} = \delta_{c2'}/H_{2'}$, where $H_{2'} = M_{c2'}/Q_{c2'}$, $\delta_{c2'} = Q_{c2'}/k_{2'}$ and $k_{2'} = 3EI_{ij,eff}\big/H_{2'}^3$. Moreover, at the base of the column jk, its chord rotation is $\theta_{3'} = \delta_{c3'}/H_{3'}$, where, $H_{3'} = M_{yc3'}/Q_{c3'}$, $\delta_{c3'} = Q_{c3'}/k_{3'}$ and $k_{3'} = 3EI_{jk,eff}\big/H_{3'}^3$. The total seismic demand rotation θ_{tot} of the joint (j) is given:

$$\theta_{tot} = \theta_{ij'} - \theta_{2'} = \theta_{jk'} - \theta_{3'} \tag{49}$$

or more accuracy:

$$\theta_{tot} = \frac{(\theta_{ij'} - \theta_{2'}) + (\theta_{jk'} - \theta_{3'})}{2} \tag{50}$$

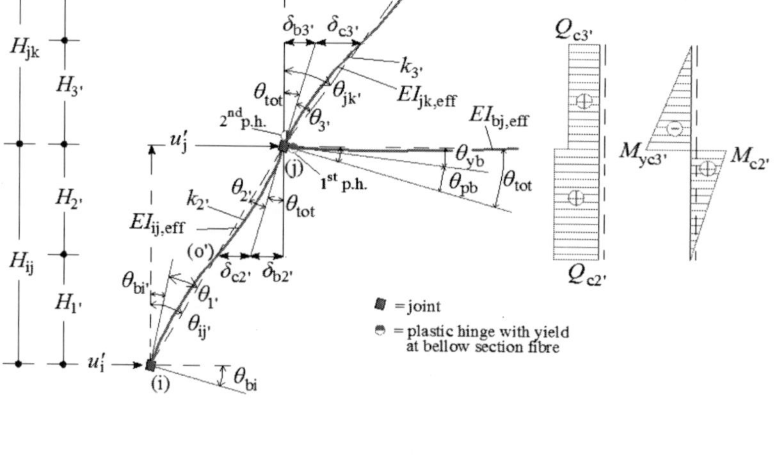

Figure 24. At exterior floor joint, where the 2nd plastic hinge is appeared at the base of the column above the joint (j).

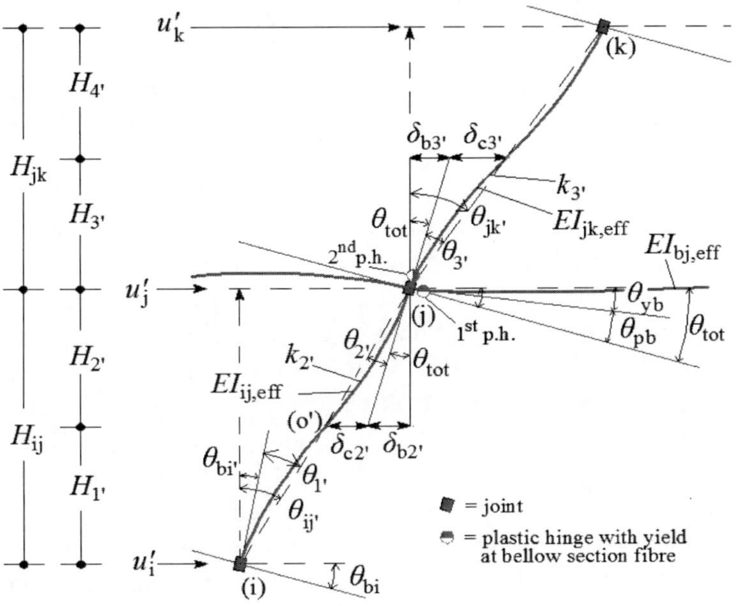

Figure 25. At interior floor joint, where the 2nd plastic hinge is appeared at the base of the column above the joint (j).

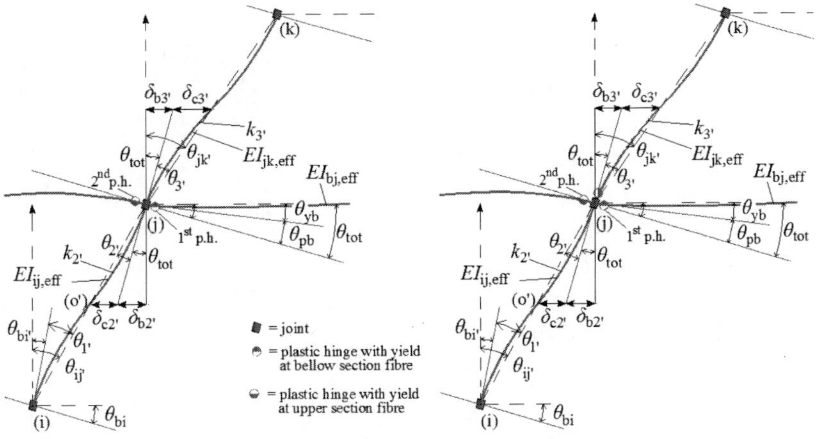

Figure 26. At interior floor joint, where the 2nd plastic hinge is appeared at the other beam-end.

Thus, just as the 2nd plastic hinge is appeared, the plastic beam-chord rotation θ_{pb} of the 1st plastic hinge is geometrically calculated:

$$\theta_{pb} = \theta_{tot} - \theta_{yb} = \frac{(\theta_{ij'} - \theta_{2'}) + (\theta_{jk'} - \theta_{3'})}{2} - \frac{(\theta_{ij} - \theta_2) + (\theta_{jk} - \theta_3)}{2} \quad (51)$$

It is worthy to note that each other case of plastic hinge is reduced at one of the abovementioned known cases.

4. Discussion

In the present Chapter, the issue of the right modelling of the nonlinear characteristics of structural r/c elements has been fully presented, where all the necessary information according to Eurocode EN 1998-3 in combination with the newly Hellenic Code of Retrofitting of Reinforced Concrete Buildings (KANEPE 2012) have been given in details. The point of the available chord rotations at yield and ultimate state of the structural r/c members using suitable stress-strain diagrams for confined and unconfined concrete, as well as for the cold worked steel has been analytically developed. In addition, the points of the available cyclic shear strength of r/c members, the effective flexural stiffness of r/c member sections, as well as the final Moment-Chord Rotation of r/c members have been presented. Moreover, the practical rational modelling of the masonry infill walls for the daily needs of r/c building design has been given. Furthermore, the calculation way of seismic demand chord rotations (elastic part and plastic one) of the structural r/c members without use springs at their ends has been analytically presented. Final, a numerical example of a r/c cantilever has been analytically given for illustrative reasons, because its seismic behavior is always the theoretical base for the completely modelling of the r/c structure according to both abovementioned contemporary seismic Codes.

References

Applied Technology Council. (1996). Seismic Evaluation and Retrofit of Concrete Buildings, Volume 1, *ATC-40 Report, Applied Technology Council,* Redwood City, California.

Applied Technology Council. (2007). *Recommended Methodology for Quantification of Building System Performance and Response Parameters.* ATC-63 90% Draft, Redwood City, CA, 2007.

Arias A. (1970). A measure of earthquake intensity. in R.J. Hansen, ed. *Seismic Design for Nuclear Power Plants*, MIT Press, Cambridge, Massachusetts: 438-483.

Asteris P.G. (2003). Lateral stiffness of brick masonry infilled plane frames. *Journal Struct. Eng.*, ASCE, 129(8): 1071-1079.

Asteris P.G. (2005). Closure to "Lateral Stiffness of Brick Masonry Infilled Plane Frames" by P.G. Asteris. *Journal of Structural Engineering*; American Society of Civil Engineers (ASCE), Vol. 131, No.3: 523-524

Asteris P.G. (2008). Finite Element Micro-Modelling of Infilled Frames. *Electronic Journal of Structural Engineering,* 8: 1-11.

Asteris P.G, Chrysostomou C., Giannopoulos I. and Smyrou E. (2011a). Masonry infilled reinforced concrete frames with openings. *Proceedings of 3rd International Conference on Computational Methods in Structural Dynamics and Earthquake Engineering* (COMPDYN 2011), 26-28 May, Corfu, Greece.

Asteris P.G., Antoniou S.T., Sophianopoulos D.S. and Chrysostomou C.Z. (2011b). Mathematical Macro-Modelling of Infilled Frames: State-of-the-Art, *Journal Of*

Structural Engineering; American Society of Civil Engineers (ASCE), doi:10.1061/(ASCE)ST.1943-541X.0000384: 1508-1517.

Asteris P.G., Kakaletsis D.J., Chrysostomou C.Z. and Smyrou E.E. (2011c). Failure Modes of Infilled Frames. *Electronic Journal of Structural Engineering*, Vol.11(1):11-20.

Asteris P.G., Giannopoulos I. and Chrysostomou C., (2012). Modelling of Infilled Frames with Openings. Special Issue on *"Advances in Infilled Framed Structures: Experimental and Modelling Aspects"*, vol.6, *The Open Construction and Building Technology Journal,* HT-BO-TOBCTJ-0006.

Athanatopoulou AM. (2005). Critical orientation of three correlated seismic components. *Journal Engineering Structures*, 27, 2, January: 301-312.

Banon H., Biggs J.M., and Irvine H.M. (1981). Seismic damage in reinforced concrete frames. *Journal of Structural Engineering*, ASCE, vol. 107, no, 9: 1713–29.

Chopra A.K. and Goel R.K. (1999). Capacity-demand-diagram methods for estimating seismic deformation of inelastic structures: SDF systems, *Report PEER-1999/02,* Pacific Earthquake Engineering Research Center, University of California, Berkeley, CA.

Eom T-S., Park H-G.. and Hong S-G. (2012). Evaluation of plastic Rotation Demands for Earthquake Design of Reinforced Concrete Beams. *ACI Structural Journal*, Technical paper, Title no. 109-S12, v.109, No1: 113-121.

EN 1992-1-1. (2004). *Eurocode 2: Design of concrete structures- Part 1-1: General rules and rules for buildings*. European Committee for Standardization, Brussels.

EN 1998-1. (2004). *Eurocode 8: Design of structures for earthquake resistance – Part 1: General rules, seismic actions and rules for buildings*. European Committee for Standardization, Brussels.

EN 1998-3. (2005). *Eurocode 8: Design of structures for earthquake resistance – Part 3: Assessment and retrofitting of building*. European Committee for Standardization, Brussels.

European Norm 12390. (2008). *Testing hardened concrete* (eight parts). European Committee for Standardization, Brussels.

Fajfar P. and Fischinger M. (1987a). N2 - a method for nonlinear seismic analysis of regular buildings. *Proceedings of 9th World Conf. Earthquake Engin.*, Tokyo, Kyoto, vol. 5: 111-116.

Fajfar P. and Fischinger M. (1987b). Nonlinear Seismic Analysis of RC Buildings: Implications of a Case Study. *European Earthquake Engineering*, vol. 1: 31-43.

Fajfar P. and Gaspersic P. (1996). The N2 Method for the Seismic Damage Analysis of RC Buildings. *Earthq. Engin.and Struct. Dynamics*, vol. 25: 31-46.

Fajfar, P. (1999). Capacity spectrum method based on inelastic demand spectra. *Earthquake Engineering and Structural Dynamics,* 28: 979-993.

Fajfar P. (2000). A Nonlinear Analysis Method for Performance-Based Seismic Design. *Earthquake Spectra*, vol.6, No3: 573-592.

Federal Emergency Management Agency. (1997a). *NEHRP Guidelines for the Seismic Rehabilitation of Buildings*. FEMA 273, Applied Technology Council, Washington, D.C.

Federal Emergency Management Agency. (1997b). *NEHRP Commentary on the Guidelines for the Seismic Rehabilitation of Buildings*. FEMA-274, Applied Technology Council, Washington, DC.

Federal Emergency Management Agency. (1998). *Evaluation of earthquake damaged concrete and masonry wall buildings: basic procedures manual*. FEMA-306. Applied Technology Council, Washington, DC.

Federal Emergency Management Agency (2009a). *Effects of Strength and Stiffness Degradation on Seismic Response*. FEMA P-440A. Prepared by the Applied Technology Council for the Federal Emergency Management Agency, Washington, D.C.

Federal Emergency Management Agency. (2009b). *Quantification of Building Seismic Performance Factors*, FEMA P-695. Prepared by Applied Technology Council for the Federal Emergency Management Agency, Washington, D.C.

Giberson M.F. (1967). *The response of* nonlinear *multi-story structures subjected to earthquake excitation.* Ph.D. Dissertation, Cal-Tech, Pasadena, California.

Goel R.K., and Chopra A.K.. (2004). Evaluation of Modal and FEMA Pushover Analyses: SAG Buildings. *Earthquake Spectra*, vol.20, No 1: 225-254.

Gupta A., and Krawinkler H. (1999). Seismic demands for performance evaluation of steel moment resisting frame structures (SAC Task 5.4.3). *Report No. 132*, John A. Blume Earthq. Engineering Center, Stanford University, Stanford, CA.

Hognestad E. (1951). A Study of Combined Bending and Axial Load in Reinforced concrete Members. *Bulletin Series No. 399, Engineering Experiment Station*, University of Illinois, Urbana, USA.

Kanaan A.E, Powell G.H. (1975). DRAIN-2D: *A General Purpose Computer Program for Dynamic Analysis of Inelastic Plane Structures*. Rep.EERC-73/6 and EERC-73/22, April 1973 and revised Aug.1975).

KANEPE (2012). *Hellenic Code of Retrofitting of Reinforced Concrete Buildings*. Organization of Seismic Design and Protection (OASP). FEK 42/b/January 20, 2012. Hellenic Ministry of Infrastructure, Transport and Network, 2012. (in Greek).

Krawinkler H, Nassar AA. (1992). Seismic design based on ductility and cumulative damage demands and capacities. *Nonlinear Seismic Analysis and Design of Reinforced Concrete Buildings*, P. Fajfar and H.Krawinkler, Eds, Elsevier Applied Science, New York, 1992.

Litton R W. (1975). *A contribution to the analysis of concrete under cyclic loading*. Ph.D. Dissertation, Dept. of Civil Engin., University of California, Berkeley.

Mainstone R.J. and Weeks G.A. (1970). The influence of Bounding Frame on the Racking Stiffness and Strength of Brick Walls, *Proceedings of the 2nd International Brick Masonry Conference,* Building Research Establishment, Watford, England: 165-171.

Mainstone R.J. (1974). Supplementary note on the stiffness and strengths of infilled frames, *Current Paper CP 13/74*, Building Research Station, Garston, Watford, U.K.

Makarios T. (2005). Optimum definition of equivalent nonlinear SDF system in pushover procedure of multistory r/c frames. *Engineering Structures Journal* v.27, 5: 814-825.

Makarios T. (2009). Equivalent Nonlinear SDF system of spatial asymmetric multistory buildings in pushover procedure. Theory and Applications. *Journal of the Structural Design of Tall and Special Buildings,* vol. 18, no. 7: 729-763.

Makarios T. (2010). Seismic nonlinear static new method of spatial asymmetric multi-storey r/c buildings. *Journal of the Structural Design of Tall and Special Buildings*. DOI: 10.1002/tal.640 (in press).

Makarios T. (2012a). The equivalent nonlinear single degree of freedom system of asymmetric multi-storey buildings in seismic static pushover analysis. CHAPTER of

Book "*Earthquake Research and Analysis* / Book 4. Book edited by Prof. Abbas Moustafa. INTECH, Open Access Publisher, ISBN 979-953-307-680-4.

Makarios T. (2012b). Evaluating of the effective seismic amplification factor on a probabilistic basis". *Structural Engineering Mechanics. An international Journal,* Vol. 42, No. 1: 121-129.

Makarios T. (2012c). Peak Ground Acceleration Functions of Mean Return Period for Known Reliability Index of Hellenic Design Earthquakes". *Journal of Earthquake Engineering,* (accepted for publication, in press).

Makarios T. and Asteris P. (2012). Numerical investigation of seismic behavior of spatial asymmetric multi-storey reinforced concrete buildings with masonry infill walls. Special Issue on "Advances in Infilled Framed structures: Experimental and Modelling Aspects". vol.6, *The Open Construction and Building Technology Journal,* HT-BO-TOBCTJ-0006.

Panagiotakos T. and Fardis M. (1999). Estimation of inelastic deformation demands in multistory rc buildings. *Journal of Earthquake Engineering and Structural Dynamics*, 28: 501-528.

Panagiotakos T. and Fardis M., (2001). Deformations of reinforced concrete members at yielding and ultimate. *ACI Structural Journal*, Technical paper, title no.98-S13 March - April, 98, No2: 135-148.

Qi X. Moehle J.P. (1991). *Displacement design approach for reinforced concrete structures subjected to earthquakes.* Report No. UCB/EERC-91/02, Univercity of California, Berkeley.

Rodriguez M. (1994). A measure of the capacity of earthquake ground motions to damage structures. *Earthquake Engin. Struct. Dynam.* 23: 627-643.

Saiidi M. and Sozen M.A. (1981). Simple Nonlinear Seismic Analysis of r/c Structures. *Journal of the Structural Division*, ASCE; 107: 937-952.

Uang C., Bertero V.V. (1990). Evaluation of Seismic Energy in Structures. *Earthq. Eng. and Struct. Dynam.*; 19: 77-90.

XTRACT. v.3.0.8. (2007). *Cross-sectional X sTRuctural Analysis of ComponenTs.* Imbsen Software System. 9912 Business Park Drive, Suite 130, Sacramento CA 95827.

PhD. Triantafyllos K. Makarios, Civil Engineer, speciality Earthquake Engineering, Senior Researcher of Institute of Engineering Seismology and Earthquake Engineering and Lecturer of Department of Civil Engineering, Aristotle University, Thessaloniki, Greece.

Born on November 8, 1964.

In: Focus on Nonlinear Analysis Research
Editors: Gualtiero Padovani and Manuel Occhino
ISBN: 978-1-62417-350-9
© 2013 Nova Science Publishers, Inc.

Chapter 2

AUTOMATIC ESTIMATION OF THE APNEA-HYPOPNEA INDEX FROM OXYGEN SATURATION RECORDINGS USING NON-LINEAR REGRESSION ANALYSIS

J. Víctor Marcos**, Roberto Hornero, Daniel Álvarez, Gonzalo C. Gutiérrez and F. Del Campo
Biomedical Engineering Group, E.T.S. Ingenieros de Telecomunicación, University of Valladolid, Spain

ABSTRACT

Generalized regression neural networks (GRNN) represent a powerful tool for modeling non-linear relationships between variables. In this study, a GRNN-based model is developed to assist in the diagnosis of sleep apnea-hypopnea syndrome (SAHS) from nocturnal oxygen saturation (SaO_2) recordings. SAHS has been pointed out as a major cause of traffic and industrial accidents and long-term effects can lead to severe cardiovascular diseases. Nowadays, nocturnal polysomnography (PSG) is the gold-standard for SAHS diagnosis. PSG must be performed in a special sleep unit and under supervision of a trained technician. Different physiological signals recorded during a complete night must be analyzed by an expert to obtain the apnea-hypopnea index (AHI). It measures the ratio of apneas and hypopneas per hour of sleep and determines SAHS severity. Despite its high diagnostic performance, PSG has several drawbacks since it is complex, costly and time-consuming. Thus, simplified diagnostic techniques are desirable. This study proposes analyzing SaO_2 signals recorded through nocturnal pulse oximetry to assist in SAHS diagnosis. The aim is to obtain a reliable estimation of the AHI from automatic analysis of these recordings. A pattern recognition approach based on GRNN was used to implement our diagnostic method. The proposed methodology comprises three different stages: 1) signal preprocessing, 2) feature extraction and 3) pattern analysis. In the first one, signal artifacts are removed. Subsequently, a total of 14 time-domain and frequency-domain features were extracted from SaO_2 recordings. In the

* AUTHOR'S ADDRESS: J. Víctor Marcos Biomedical Engineering Group, E.T.S. Ingenieros de Telecomunicación, University of Valladolid, Paseo Belén, 15, 47011 - Valladolid (Spain). Telephone: +34 983184716. Fax: +34 983423667, E-mail: jvmarcos@gmail.com, Web-site: www.gib.tel.uva.es.

third stage, these features were used as inputs to a GRNN model in order to estimate the AHI. A database composed of 240 SaO_2 signals was available for this study. It was randomly divided into a training set (40%) with 96 SaO_2 signals and a test set (60%) with 144 SaO_2 signals. We found that our GRNN-based model achieved an intraclass correlation coefficient (ICC) of 0.89 on the test set, showing high reliability in the estimation of the AHI. According to our results, the proposed GRNN model represents a valuable screening tool for SAHS. It could be used in clinical practice in order to reduce the number of required PSG tests.

Keywords: Biomedical signal processing, nonlinear regression, sleep apnea-hypopnea syndrome, oxygen saturation

1. INTRODUCTION

Nowadays, nocturnal polysomnography (PSG) is required to obtain a reliable diagnosis about SAHS. PSG is performed in a special sleep unit and enables the detection of apneas, hypopneas and arousals. It includes the acquisition of several recordings, such as electroencephalogram (EEG), electrocardiogram (ECG) or arterial blood oxygen saturation (SaO_2), in addition to a range of other variables required for the complete description of sleep (Qureshi and Ballard 2003). The PSG study is supervised by qualified personnel (technicians and physicians) able to perform, monitor and interpret the examination (Gagnadoux et al. 2002). Once the test has been completed, a medical expert must manually analyze all the physiological recordings to provide a definitive diagnosis. Every disordered-breathing event is identified from the recorded data. According to the American Academy of Sleep Medicine, an apnea is considered as a cessation of airflow for at least 10 seconds (Kushida et al. 2005). On the other hand, a hypopnea event is considered as a reduction in airflow of at least 50% that is associated with an oxygen desaturation greater than 4% (Kushida et al. 2005). The frequency of apneas and hypopneas per hour of sleep is expressed as the apnea-hypopnea index (AHI), which is used to characterize the severity of SAHS. Usually, an AHI between 5 h^{-1} and 15 h^{-1} is used to determine a positive diagnosis (Caples et al. 2005).

Despite its diagnostic reliability, PSG presents some drawbacks. The sleep laboratory is an artificial environment and some patients present a disturbed sleep pattern due to the foreign setting. PSG is complex and time-consuming due to the great quantity of information that must be processed by the sleep specialist. In addition, it is an expensive procedure since specific equipment and personnel are required (Bennet and Kinnear 1999). Finally, the reception capacities of currently existing sleep laboratories are overwhelmed. The demand for this type of studies is progressively growing due to a better understanding of the disease and its associated risks. In contrast, the infrastructure to support it seems to be insufficient (Flemons et al. 2004). It is suspected that a high percentage of patients suffering from SAHS remain undiagnosed. The study by Young et al. (1997) found that 82% of men and 93% of women with moderate to severe SAHS were not diagnosed. Early treatment is required in order to prevent other health complications. Flemons et al. (2004) evaluated the capacity of the sleep studies of five developed countries: United Kingdom, Belgium, Australia, United States and Canada. The average waiting time from the referral to a specialist to the beginning of treatment was estimated. It ranged from 2 months in Belgium up to 7-60 months in the

United Kingdom. The number of beds prepared to perform PSG studies per 100000 inhabitants was also computed. It was 0.3 in the United Kingdom while the rate was around 1.4 in the other countries.

These statistics reveal the need for improving the protocol for SAHS diagnosis. Increasing the capacity for PSG studies by developing the existing sleep laboratories or creating new ones is required. However, it seems to be slow and an important investment on resources is needed. Other alternatives based on technological advances can be found. For instance, home unattended and hospital telemonitored PSG have been proposed (Gagnadoux et al. 2002). However, both are still complex, time-consuming and expensive. Research work related to the development of simplified diagnostic techniques has increased in the last years. The aim is to provide an accurate tool for SAHS detection while avoiding the difficulties of PSG. In particular, the usefulness of nocturnal oximetry for SAHS screening has been widely assessed. Oximetry recordings reflect respiratory dynamics during sleep. Pulse oximetry has the advantages of being readily available, relatively inexpensive and can be performed at home (Bennet and Kinnear 1999). Therefore, SaO_2 analysis has been suggested as a suitable approach to perform SAHS diagnosis and to reduce the need for PSG studies (Chiner et al. 1999; Caples et al. 2005).

Pulse oximetry is an essential tool in sleep medicine to determine a patient's cardiorespiratory stability. Respiratory disturbances such as apneas and hypopneas are reflected in the SaO_2 profile. The saturation value decreases due to reduction of airflow. Subsequently, it returns to its baseline level when respiration is restored. Therefore, SaO_2 recordings from SAHS-positive patients are characterized by marked fluctuations due to desaturation events, which reflect unstable ventilation (Bloch, 2003). In contrast, healthy respiratory patterns tend to present a near-constant SaO_2 waveform around 96% (Netzer et al. 2001). To illustrate this, the SaO_2 signals corresponding to a control subject (AHI = 1.4 h^{-1}), a doubtful control subject (AHI = 8.1 h^{-1}), a doubtful SAHS-positive patient (AHI = 13.1 h^{-1}) and a SAHS-positive patient (AHI = 66.2 h^{-1}) are depicted in Figure 1. As can be observed, a more unstable behavior is reflected in patients with a higher AHI value. However, it is difficult to discriminate signals from doubtful subjects.

Traditionally, analysis of nocturnal oximetry was focused on visual inspection of tracings for identification of desaturation events, baseline and minimal values, and patterns of fluctuation in oxygen saturation (Bloch, 2003). However, it is tedious and subjective. A manual score must be provided for recordings extending over several hours and interpretation may differ among experts. Indeed, there is no universally accepted definition of an oxygen desaturation in sleep-disordered breathing (a decrease greater than 4% is usually applied) (Netzer et al. 2001). Several desaturation indices were proposed to perform automated and objective analysis of oxymetry signals. They reflect the variability of oxygen saturation. The most common indices are the oxygen desaturation index over 3% (ODI3) and 4% (ODI4) and the cumulative time spent below a given level of saturation, which is usually fixed at 90% (CT90) (Bloch, 2003). Their diagnostic utility has been evaluated in previous studies (Netzer et al. 2001). The reported results show a substantial difference between sensitivity and specificity. The former varied from 88 to 98% and the latter from 40 to 88%, when an AHI of 10 h^{-1} or 15 h^{-1} measured during PSG was defined as the upper limit of normal (Bloch, 2003). Thus, none of these indices can be pointed out as a reliable measurement to perform SAHS diagnosis.

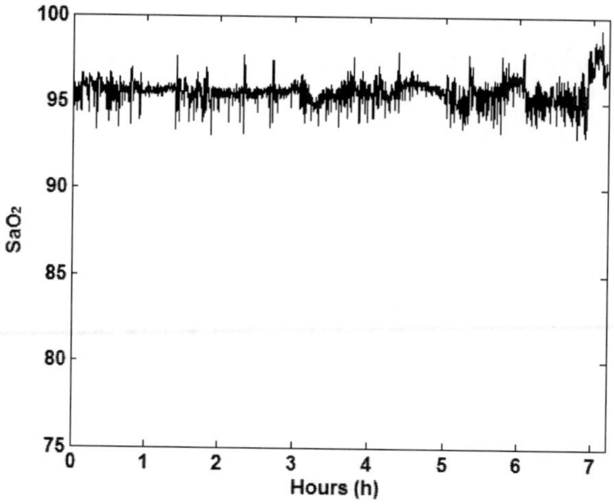

(a)

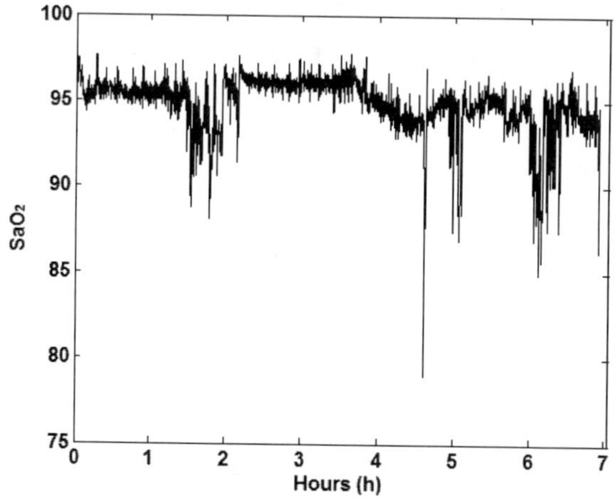

(b)

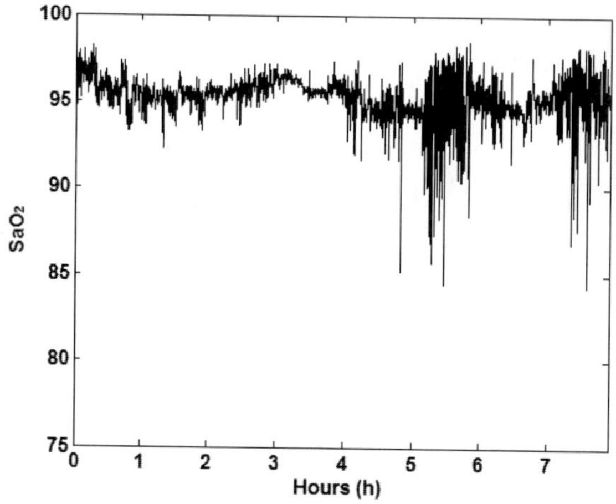

(c)

Figure 1. (Continued).

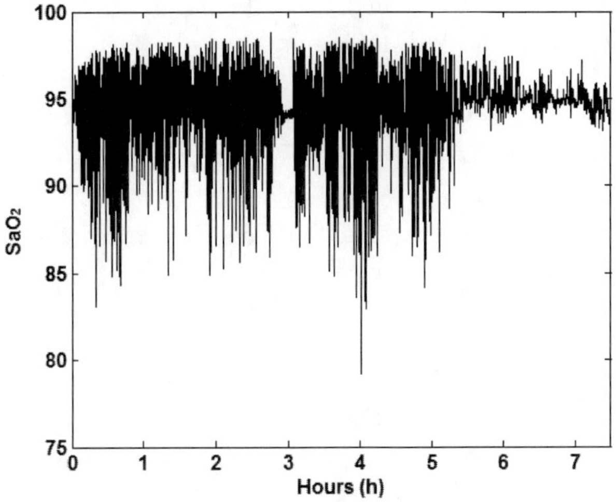

(d)

Figure 1. Examples of SaO_2 signals corresponding to (a) a SAHS-negative subject (AHI = 1.4 h^{-1}), (b) a doubtful SAHS-negative subject (AHI = 8.1 h^{-1}), (c) a doubtful SAHS-positive subject (AHI = 13.1 h^{-1}) and (d) a SAHS-positive subject (AHI = 66.2 h^{-1}).

In this study, it is hypothesized that the diagnostic capability of oximetry recordings can be enhanced by using advanced signal processing techniques. According to the dynamic behavior of SaO_2 signals, time and frequency analysis may be appropriate to identify features reflecting the presence of SAHS and its degree of severity. Previous studies have shown that both spectral (Zamarrón et al. 2003) and non-linear (Álvarez et al. 2006; Del Campo et al. 2006; Hornero et al. 2007) features from oximetry data provide useful information about SAHS. It is proposed to apply multivariate regression techniques to combine these features in order to obtain an accurate estimation of the AHI. Several preceding studies presented diverse assistant methods for SAHS diagnosis based on the analysis of multivariate patterns. For instance, discriminant analysis (Marcos et al. 2009; Marcos et al. 2010b), multilayer perceptron (MLP) neural networks (Marcos et al. 2008b; Marcos et al. 2010a; Marcos et al. 2012) and radial basis function (RBF) neural networks (Marcos et al. 2008a) have been evaluated. A methodology composed of three different stages is proposed in the present research for automatic computation of the AHI from SaO_2 data. In the first one, oximetry recordings are preprocessed in order to suppress artifacts. Subsequently, the second stage maps the SaO_2 signal onto a set of time-domain and frequency-domain features. In the third stage, regression analysis is performed to model the dependence of the AHI on these features. Generalized regression neural networks (GRNN) represent a common procedure for this task and were chosen for regression analysis in this study. They are based on the estimation of the regression function using the Parzen window method (Specht, 1991). Previously, these networks have been applied to different regression problems such as time-series forecasting (Yan, 2012), prediction of water consumption (Firat et al. 2010) or fault diagnosis (Lee et al. 2004). Additionally, GRNN algorithms have been used for SAHS detection from clinical data (Kirby et al. 1999).

The aim of this research is to develop an automated method to assist in SAHS diagnosis from SaO_2 analysis. We seek to obtain a reliable estimation of the AHI without the requirement of a complete sleep study. In order to do this, a detailed analysis of SaO_2 data is proposed by using multivariate analysis techniques.

2. Subjects and Signals

The population available for this study was composed of 240 subjects suspected of suffering from SAHS. There was no distinction about the type of apneas (obstructive, central or mixed) for recruitment of subjects. All of them presented typical symptoms such as daytime sleepiness, loud snoring, nocturnal choking and awakenings or apnea events reported by the subject or a bedmate. The data set consisted of 80 SAHS-negative and 160 SAHS-positive subjects. The SAHS-negative group was composed of 52 men and 28 women. Their mean ± standard deviation age, BMI and AHI were 47.23 ± 12.15 years, 27.83 ± 3.84 kg/m^2 and 3.94 ± 2.40 h^{-1}, respectively. On the other hand, a total of 134 men and 26 women formed the SAHS-positive group. Their mean age and BMI were 54.78 ± 13.79 years and 30.83 ± 4.34 kg/m^2, respectively. The average AHI value in this group was 36.63 ± 25.73 h^{-1}. The percentage of males was significantly higher in the population of patients. As can be observed, the proportion of SAHS-positive patients is substantially greater than that of controls since subjects were referred to sleep study because of their symptoms and suspicions about SAHS.

The initial database was divided into training and test sets in order to provide an unbiased evaluation of the regression algorithms. Approximately, 40% of the subjects under study were assigned to the training set whereas the remaining group was used for testing. Several random divisions of the initial population were carried out. Ideally, both sets should present similar mean values for age, percentage of males, BMI and recording time. Thus, the partition with the best balance for these variables was selected. The training set was allocated for algorithm optimization. Usually, pattern recognition techniques require some adaptive parameters to be optimized according to the statistical properties of the problem. The training set, composed of a set of representative examples of the problem, was used for this purpose. In this study, the training set consisted of 96 subjects (32 SAHS-negative and 64 SAHS-positive). Once the optimization process is completed, the trained algorithm is ready to be used. A test set composed of previously unseen samples is required to objectively assess the performance of the algorithm in the proposed problem. In this study, the oximetry signals from 144 subjects (48 SAHS-negative and 96 SAHS-positive) composed the test set. These signals were not previously included in the data set used for training. Table 1 summarizes clinical and demographic data for the whole data set as well as for training and test sets separately.

Overnight conventional PSG was performed from midnight to 08:00 AM in the Sleep Unit of Hospital Universitario Pío del Río Hortega of Valladolid, Spain. The Review Board on Human Studies approved the protocol and each subject gave his informed consent to participate in the study. Patients were continuously monitored using a polysomnograph (Alice 5, Respironics, Philips Healthcare, The Netherlands). A full description of the equipment and its technical specifications can be found in the Philips Respironics website (Philips Respironics). PSG studies included EEG, EOG, chin EMG, airflow (thermal and pressure flow), ECG, measurement of body position and oximetry. Signals are sent from patient interface to headbox through sensor cables.

Data was sampled and sent to the base station where it is stored on a disk or sent through an ethernet connection to a computer. A sleep specialist analyzed these recordings according to the system proposed by Rechtschaffen and Kales (Rechtschaffen and Kales 1968).

The derived AHI was used to provide a reliable diagnosis for each subject. It served as the reference for validating the results obtained in this study. The methods proposed for automated SAHS diagnosis are based on SaO_2 analysis. Oximetry signals obtained through PSG composed the database available for this study. The polysomnograph equipment included a Nonin PureSAT pulse oximeter (Nonin Medical Inc., USA) with 3 seconds or faster averaging interval at a minimum heart rate of 60 beats per minute or greater. A finger probe was used to record these signals at a sampling frequency of 1 Hz. The recordings were saved to separate files to be off-line processed.

3. METHODS

A methodology based on three stages was designed to implement the AHI estimator using generalized regression neural networks (GRNN) models. In the first one, signal preprocessing is performed in order to remove artifacts. A feature pattern is extracted from SaO_2 data in the second stage by means of both time and frequency domain analysis. Finally, a GRNN model is used in the last stage to provide an estimation of the AHI from the extracted pattern.

3.1. Signal Preprocessing

One of the limitations of overnight pulse oximetry is given by motion artifacts caused by any movement of the probe on the extremity. Pulse oximetry is especially sensitive to this problem as the energy absorption due to arterial pulsation is only 1-2% of the total absorption (Moyle, 1996). Drops to zero observed in SaO_2 signals reflect artifacts occurring during sleep. These samples should be removed before performing analysis of oximetry data. However, removing zero samples may not be enough since some signals epochs could be affected by frequent drops. In this case, non-zero samples between consecutive drops do not provide a reliable measure of SaO_2 and the complete signal epoch should be removed.

A preprocessing stage was applied to SaO_2 signals in order to remove these artifacts. Magalang et al. (2003) proposed to eliminate all changes of SaO_2 signals between consecutive sampling intervals greater than 4%/s and any SaO_2 sample lower than 20%. These criteria were taken into account for implementing our preprocessing algorithm. It evaluates each sample in the recording. The aim was to detect zero-samples and markedly affected signal epochs. The preprocessing algorithm involves the following steps:

1. Samples corresponding to the first 5 minutes of the signal are discarded.
2. A moving window with a length of 10 minutes is used to define a reference level. It is computed as the median value of the samples in this window. Initially, it is empty and new valid samples are progressively included in the window.

3. A pointer is used to analyze each sample of the signal. Three different states are defined:
   3.1. If the sample is distinct to zero, it is accepted as a valid sample. The moving window is updated and the pointer moves towards the next sample.
   3.2. If the sample is equal to zero, it is discarded and the pointer moves to the next sample.
   3.3. If a non-zero sample is found after a zero-sample, it will be accepted if the two following conditions are met:
      - The absolute difference between the value of the sample and the reference level obtained from the moving window should be less than 5% of the latter.
      - The next one-minute signal epoch (counted from the sample that met the previous condition) should present non-zero values.
4. If these two conditions are satisfied, the one-minute signal epoch after the non-zero sample is marked as valid, the moving window is updated and the pointer is placed in the next sample after this epoch. If there are zero values, the pointer is placed in the next sample after the last zero value in the one-minute epoch. The sample contained by the pointer is evaluated. The algorithm will take one of the three possible states described in step 3.

Table 1. Clinical and demographic features for the available database of subjects as well as for training and test sets. BMI: body mass index; AHI: apnoea-hypopnoea index

ALL SUBJECTS			
	All	SAHS-positive	SAHS-negative
Subjects	240	160	80
Age (years)	52.26 ± 13.71	54.78 ± 13.79	47.23 ± 12.15
Males (%)	77.50	83.75	65.00
BMI (kg/m^2)	29.83 ± 4.38	30.83 ± 4.34	27.83 ± 3.84
Recording Time (h)	7.25 ± 0.55	7.23 ± 0.64	7.28 ± 0.31
AHI (h^{-1})	25.73 ± 26.09	36.63 ± 25.73	3.94 ± 2.40
TRAINING SET			
	All	SAHS-positive	SAHS-negative
Subjects	96	64	32
Age (years)	52.35 ± 13.76	54.88 ± 14.53	47.31 ± 10.59
Males (%)	77.08	84.38	62.50
BMI (kg/m^2)	29.83 ± 4.17	30.61 ± 3.86	28.27 ± 4.38
Recording Time (h)	7.25 ± 0.33	7.25 ± 0.35	7.25 ± 0.29
AHI (h^{-1})	24.75 ± 25.19	35.01 ± 25.16	4.23 ± 2.22
TEST SET			
	All	SAHS-positive	SAHS-negative
Subjects	144	96	48
Age (years)	52.19 ± 13.73	54.71 ± 13.35	47.17 ± 13.20
Males (%)	77.78	83.33	66.67
BMI (kg/m^2)	29.83 ± 4.53	30.98 ± 4.65	27.54 ± 3.26
Recording Time (h)	7.24 ± 0.66	7.22 ± 0.78	7.30 ± 0.33
AHI (h^{-1})	26.39 ± 26.74	37.71 ± 26.17	3.75 ± 2.51

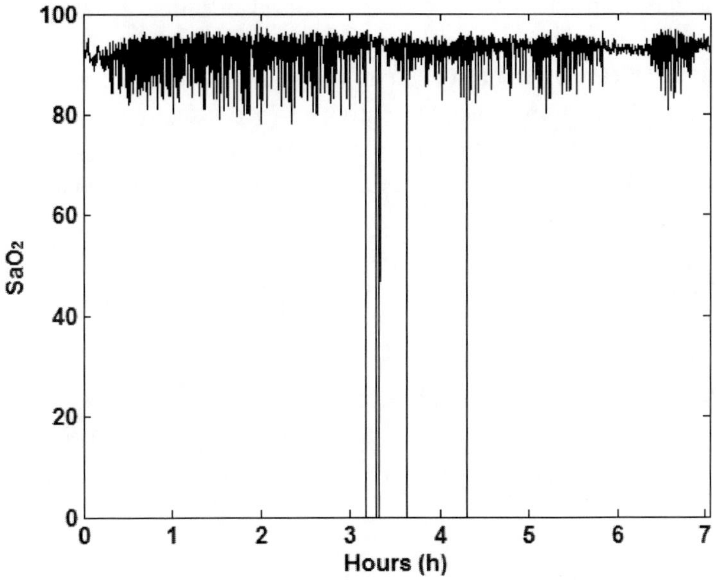

(a)

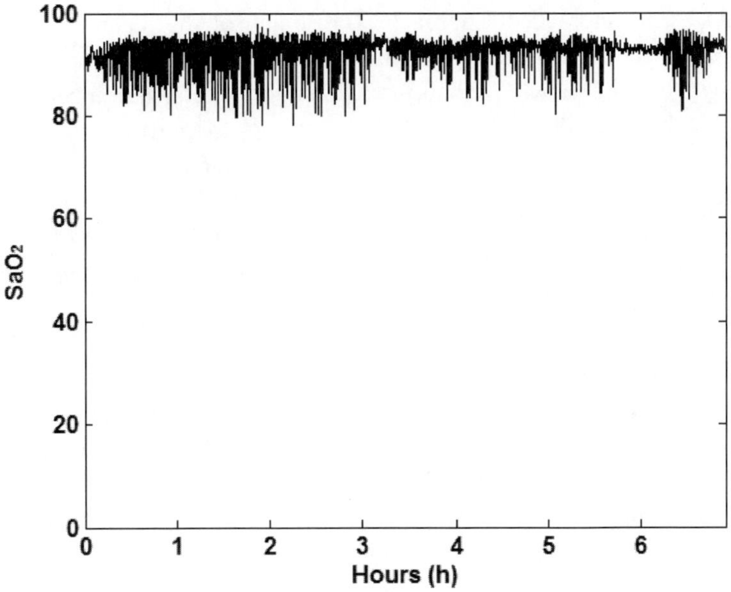

(b)

Figure 2. Example of a SaO_2 recording (a) before preprocessing and (b) after artefact removal using the proposed algorithm. As can be observed, drops to zero and subsequent abnormal values of saturation are removed.

Each time the algorithm accepts a signal epoch in state 3.3, the new segment, which was preceded by zero samples, must be assembled with the previous part of the signal. The following criterion was used:

5. A maximum step parameter of 4% (expressed in units of SaO_2) is considered for joining two different signals epochs (Magalang et al. 2003). The difference between extreme samples, i.e., the last sample identified as valid and the first sample of the new signal epoch is computed. It is divided by the step parameter and the obtained result is rounded towards minus infinity. The result provides the number of new samples to be included between both extreme samples. Interpolation techniques are used to compute the value of the signal in these new points.

Figure 2 shows an example of a SaO_2 recording after being processed for artifact removal. As can be observed, drops to zero and signal epochs containing repetitive frequent artifacts are automatically removed. The dynamic properties of the signal are preserved since no digital filtering is applied.

3.2. Feature Extraction

Oximetry signals were recorded during a complete night of sleep. The average recording time from all signals used in this study was 7.25 hours, which corresponds to more than 26000 samples. Using each sample in the recording as an input variable to the GRNN algorithm would be an impractical approach. The length of the recording varies from one subject to another. In addition, the input dimension would be extremely high. This would lead to an intractable system that would probably provide poor generalization because of the curse of dimensionality (Bishop, 1995). Therefore, it is necessary to transform the signal into a new representation before regression analysis. Feature extraction is required to map the SaO_2 signal into a reduced set of variables or features that summarize the information in the recording. The term feature extraction is used to denote the process of forming linear or non-linear combinations of original data to generate new variables (Bishop, 1995). Each of them represents a specific attribute of the signal.

In this study, SAHS diagnosis was modeled as a regression problem. Therefore, the extracted features should correlate with the degree of severity of SAHS to obtain a suitable AHI estimation. According to the domain used for analyzing SaO_2 signals, features were divided into two groups: time-domain and frequency-domain features. The former are related to the temporal evolution of signals. The latter refers to the study of their frequency components. Both groups of features were used to characterize the dynamic properties of SaO_2 signals regarding SAHS.

3.2.1. Time-Domain Analysis of SaO_2 Data

Time representation of oximetry signals reflects respiratory dynamics during sleep. Apnea events are characterized by a decrease in the SaO_2 value due to airway obstruction and reduced airflow. Therefore, signals from SAHS-positive subjects are usually associated to instability. They reflect continuous drops and subsequent restorations of the saturation value due to the repetition of apneas. In contrast, signals from SAHS-negative subjects tend to present a near constant saturation value around 96% (Netzer et al. 2001). In order to quantify these dynamic differences, oximetry recordings were analyzed in the time domain using conventional statistics and non-linear methods.

The statistical properties of the SaO_2 variable were analyzed. The first four standard moments of this variable were computed. To evaluate these features, the probability density function was modeled with a discrete distribution, assigning to each saturation value in the series $\mathbf{s} = \{s_1, ..., s_T\}$ the probability given by its relative frequency, i.e. the number of times that the SaO_2 value was observed out of the total number of samples given by T. The definition of these moments are given below (Jobson, 1991a):

Feature 1. First statistical moment in the time domain or mean (μ_t). It quantifies the central tendency for the distribution of SaO_2 samples and is given by:

$$\mu_t = \frac{1}{T}\sum_{i=1}^{T} s_i \tag{1}$$

Feature 2. Second statistical moment in the time domain or variance (σ_t^2). It indicates the degree of dispersion for the distribution of SaO_2 samples. The variance was computed as:

$$\sigma_t^2 = \frac{1}{(T-1)}\sum_{i=1}^{T}[s_i - \mu_t]^2 \tag{2}$$

Feature 3. Third statistical moment in the time domain or skewness (γ_t). It measures the asymmetry for the distribution of SaO_2 samples. A negative skew indicates that the tail on the left side of the distribution is longer than on the right side and most of the values lie on the right of the mean. Conversely, a positive skew is obtained for distributions with a longer tail on the right side. The definition of skewness is given by:

$$\gamma_t = \frac{\frac{1}{(T-1)}\sum_{i=1}^{T}[s_i - \mu_t]^3}{\sigma_t^3} \tag{3}$$

Feature 4. Fourth statistical moment in the time domain or kurtosis (δ_t). It evaluates the sharpness for the distribution of SaO_2 samples. Higher kurtosis reflects the existence of infrequent values that contribute to increase the variance. It was computed as:

$$\delta_t = \frac{\frac{1}{(T-1)}\sum_{i=1}^{T}[s_i - \mu_t]^4}{\sigma_t^4} \tag{4}$$

Non-linear analysis was also used to assess oximetry dynamics in the time domain. Irregularity, variability and complexity of SaO_2 recordings were evaluated by using approximate entropy (*ApEn*), central tendency measure (*CTM*) and Lempel-Ziv complexity (*LZC*), respectively. It has been shown that the measures provided by these methods correlate with the severity of SAHS (Álvarez et al. 2006; Hornero et al. 2007). A brief description of these non-linear methods used for feature extraction is provided:

Feature 5. Approximate entropy (*ApEn*). *ApEn* is a family of parameters and statistics introduced to quantify irregularity in data (Pincus, 2001). To compute *ApEn* from a series of length T, patterns $\mathrm{s}_i = \{s_i, s_{i+1}, \ldots, s_{i+m-1}\}$ composed of m consecutive samples from the original sequence s are obtained. The probability of occurrence of a given m-length pattern $C_r^m(i)$ is expressed as:

$$C_r^m(i) = \frac{N^m(i)}{T-m+1} \tag{5}$$

where $N^m(i)$ denotes the number of patterns s_j of length m to a distance less or equal than r from s_i. The *ApEn* is defined as the following ratio (Pincus, 2001):

$$ApEn(m,r) = \lim_{T\to\infty}\left[\phi^m(r) - \phi^{m^*}(r)\right] =$$
$$\lim_{T\to\infty}\left[\frac{1}{T-m+1}\sum_{i=1}^{T-m+1}\ln C_r^m(i) - \frac{1}{T-m^*+1}\sum_{i=1}^{T-m^*+1}\ln C_r^{m^*}(i)\right] \tag{6}$$

where $m^* = m + 1$ and $\phi^m(r)$ represents the average logarithmic likelihood for the m-length pattern s_i.

Feature 6. Central tendency measure (*CTM*). *CTM* provides a measure of signal variability. It is obtained from second-order difference scatter plots representing $(s_{k+2} - s_{k+1})$ vs. $(s_{k+1} - s_k)$. A circular region of radius ρ_{CTM} is defined around the origin to compute *CTM*. The number of points that fall in this region is counted and divided by the total number of points (Cohen et al. 1996):

$$CTM = \frac{\sum_{i=1}^{T-2}\delta(d_i)}{T-2} \tag{7}$$

where $\delta_k = 1$ if the kth point is inside the circle and 0 otherwise.

Feature 7. Lempel-Ziv complexity (*LZC*). *LZC* is a non-parametric measure of complexity in a one-dimensional signal. It is related to the number of distinct substrings and the rate of their recurrence along a given sequence. To compute *LZC*, the original signal s must be transformed into a two-symbol sequence $\mathrm{p} = \{p_1, \ldots, p_i, \ldots, p_T\}$. Each SaO_2 sample is compared with the median value of the samples to perform the transformation. Then, the sequence p is scanned from left to right and the complexity counter $c(T)$ is increased by one unit every time a new subsequence of consecutive characters is encountered. The value of *LZC* is given by (Lempel and Ziv 1976):

$$LZC = \frac{c(T)}{b(T)} \tag{8}$$

where $b(T)$ is a normalization factor. It is given by:

$$b(T) = \frac{T}{\log_2(T)} \tag{9}$$

3.2.2. Frequency-Domain Analysis of SaO_2 Data

Spectral analysis of SaO_2 signals reveals the influence of apnea/hypopnea events on signal dynamics. Preceding studies found that signal power associated with frequency components between 0.010 and 0.033 Hz is usually higher in subjects with increased AHI (Zamarrón et al. 2003). Apneas originate phase-lagged changes in SaO_2 signals. Their duration usually ranges from 30 seconds to 2 minutes, including the awakening response after the event. Patients suffering from SAHS may have several consecutive breathings reflecting apneas or hypopneas. Thus, the repetition of these events during these periods of sleep will be produced at a rate between 30 seconds and 2 minutes, which correspond with a frequency of 0.033 Hz and 0.010 Hz, respectively. Relevant frequency components in this range denote fluctuations in oximetry recordings due to phases with repetitive apneas. This behavior can be examined through spectral analysis of oximetry data.

The non-parametric Welch's method was used to compute the PSD of oximetry recordings (Welch, 1967). Briefly, the original recording $\mathbf{s} = \{s_1, ..., s_T\}$ is divided into M overlapping sequences of length L by applying a window function $v = \{v_1, \ldots, v_L\}$. The modified periodogram is computed for each of them by using the finite Fourier transform. It is given by:

$$\hat{S}_m(k) = \frac{1}{LU}\left|\sum_{i=0}^{L-1} s_i v_i e^{-\frac{j2\pi ki}{L}}\right|^2 \tag{10}$$

where

$$U = \frac{1}{L}\sum_{i=0}^{L-1}|v_i|^2 \tag{11}$$

Finally, the PSD estimation is obtained as the average of the periodograms:

$$PSD(f_k) \approx \hat{S}(k) = \frac{1}{M}\sum_{m=0}^{M-1}\hat{S}_m(k) \tag{12}$$

Two sets of features were derived from the PSD function (Marcos et al. 2009; Marcos et al. 2010). Initially, statistical analysis was used to characterize the PSD of SaO_2 signals. The variable representing the frequency component was modeled using the normalized PSD

(PSDn) as its probability density function. Only the positive half of the original PSD was considered. As proposed in time-domain analysis, the first four standard moments were used to characterize this variable:

Feature 8. First statistical moment in the frequency domain or mean (μ_f). It estimates the central tendency of the variable defined by the frequency component. The value of μ_f is computed as:

$$\mu_f = \sum_{f=0}^{f_s/2} PSD_n(f) f \tag{13}$$

Feature 9. Second statistical moment in the frequency domain or variance (σ_f^2). It represents the degree of dispersion of the variable defined by the frequency component. The following expression was used to compute the variance of variable f:

$$\sigma_f^2 = \sum_{f=0}^{f_s/2} PSD_n(f)(f - \mu_f)^2 \tag{14}$$

Feature 10. Third statistical moment in the frequency domain or skewness (γ_f). It measures the asymmetry of the variable defined by the frequency component. This feature was obtained as:

$$\gamma_f = \frac{\sum_{f=0}^{f_s/2} PSD_n(f)(f - \mu_f)^3}{\sigma_f^3} \tag{15}$$

Feature 11. Fourth statistical moment in the frequency domain or kurtosis (δ_f). It evaluates the sharpness of the distribution defined by the frequency component. Kurtosis was computed as:

$$\delta_f = \frac{\sum_{f=0}^{f_s/2} PSD_n(f)(f - \mu_f)^4}{\sigma_f^4} \tag{16}$$

In addition, three features were directly computed from the PSD function. The properties evaluated by them are described in the following points:

Feature 12. Total area under the PSD (S_T). It evaluates the power of the SaO_2 signal under study and is obtained as:

$$S_T = \sum_{f=0}^{f_s/2} PSD(f) \tag{17}$$

Feature 13. Area enclosed in the band of interest (S_B). This feature measures the signal power corresponding to frequencies in the band between 0.010 and 0.033 Hz. The expression to compute S_B is given by:

$$S_B = \sum_{f=0.010}^{0.033} PSD(f) \tag{18}$$

Feature 14. Peak amplitude of the PSD in the band of interest (PA). It represents the most significant frequency component contained in the band between 0.010 and 0.033 Hz. The definition of PA is given by:

$$PA = \max_{PSD(f)} \{PSD(f)\},\ f \in [0.010, 0.033](Hz) \tag{19}$$

The extracted features represent different properties of the signal under analysis. Therefore, their values may differ significantly. Simple linear rescaling is useful to avoid differences between the magnitudes of the input features. Each feature ($\tilde{x}_i^n$) was independently treated. Its mean (μ_i) and standard deviation (σ_i) were computed with respect to the training set. A set of rescaled variables were obtaining by applying the following linear transformation (Bishop, 1995):

$$x_i^n = \frac{\tilde{x}_i^n - \mu_i}{\sigma_i} \tag{20}$$

where n and i are the sample and feature indices, respectively, is the normalized value of feature i for sample n and is its corresponding raw value. The transformed variables have zero mean and unit standard deviation.

3.3. Generalized Regression Neural Networks (GRNN)

Regression analysis was carried out to approximate the functional relationship between oximetry and the AHI. Using the maximum likelihood principle, the optimum estimator y of a continuous variable t is that which minimizes the sum-of-squares error function (E_D) (Bishop, 1995):

$$E_D = \frac{1}{2}\sum_{n=1}^{N}\left[y(\mathbf{x}^n, \mathbf{w}) - t^n\right]^2 \tag{21}$$

where w represents a set of adaptive parameters or weights. By minimizing the previous error term, y represents the least-squares solution to the problem of fitting data in the training set D. This set is composed of example pairs ($\mathbf{x}^n$, t^n) given by an input pattern $\mathbf{x}^n$ and its corresponding desired response t^n. In our study, the input pattern is composed of the normalized features derived from SaO_2 data, as described in the previous stage, while the target variable is represented by the AHI value.

It can be shown that the minimum E_D is achieved by the following (optimum) estimator (Bishop, 1995):

$$y\left(\mathbf{x},\mathbf{w}^*\right)=\left\langle t|\mathbf{x}\right\rangle \tag{22}$$

where $\mathbf{w}^*$ denotes the value of the weight vector that leads to the minimum E_D. This result indicates that using a sum-of-squares error for optimizing the model $y(\mathbf{x}, \mathbf{w})$, its output approximates the conditional average of the target data. This function is known as the regression of t conditioned on x.

It is worth noting that any prior assumption about the architecture of the model was made. The expression of the regression function shows that if we could make several measurements of t for a given x, we would obtain a set of random values whose average (in the limit of an infinite sample) defines the value of the target function at point $\mathbf{x}$ (Bishop, 1995).

In this study, GRNN algorithms were used to perform regression analysis. GRNN models are based on directly computing the regression function as expressed in Eq. (22). For this purpose, the Parzen window procedure was used for probability density estimation (Parzen, 1962). The definition of the regression function is given by:

$$\left\langle t|\mathbf{x}\right\rangle=\int tp\left(t|\mathbf{x}\right)dt=\frac{\int tp\left(\mathbf{x},t\right)dt}{p\left(\mathbf{x}\right)}=\frac{\int tp\left(\mathbf{x},t\right)dt}{\int p\left(\mathbf{x},t\right)dt} \tag{23}$$

Therefore, the joint probability function of variables x and t enables the computation of the conditional probability density function $p(t|\mathbf{x})$ and its expected value.

In GRNN models, the training set D is used to estimate $p(\mathbf{x}, t)$ using the Parzen window procedure. The following approximation for the joint probability density function is considered (Specht, 1991):

$$p\left(\mathbf{x},t\right)\approx\frac{1}{\left(2\pi\right)^{(d+1)/2}\sigma_g^{(d+1)}}\frac{1}{N}\sum_{n=1}^{N}\exp\left[-\frac{\left(\mathbf{x}-\mathbf{x}^n\right)^T\left(\mathbf{x}-\mathbf{x}^n\right)}{2\sigma_g^2}\right]\exp\left[-\frac{\left(t-t^n\right)^2}{2\sigma_g^2}\right] \tag{24}$$

Substituting this approximation into the expression of the desired regression function in Eq. (23) leads to the following result:

$$\langle t|\mathbf{x}\rangle \approx \frac{\sum_{n=1}^{N} \exp\left[-\frac{\left(\mathbf{x}-\mathbf{x}^n\right)^T\left(\mathbf{x}-\mathbf{x}^n\right)}{2\sigma_g^2}\right]\int t \exp\left[-\frac{\left(t-t^n\right)^2}{2\sigma_g^2}\right]dt}{\sum_{n=1}^{N} \exp\left[-\frac{\left(\mathbf{x}-\mathbf{x}^n\right)^T\left(\mathbf{x}-\mathbf{x}^n\right)}{2\sigma_g^2}\right]\int \exp\left[-\frac{\left(t-t^n\right)^2}{2\sigma_g^2}\right]dt} \tag{25}$$

The integral in the numerator is the mean value for Gaussian variable t centered at t^n, while the integral in the denominator reduces to unity. Defining the scalar function $(\theta^n)^2$ as

$$\left(\theta^n\right)^2 = \left(\mathbf{x}-\mathbf{x}^n\right)^T\left(\mathbf{x}-\mathbf{x}^n\right) \tag{26}$$

yields the following expression:

$$\langle t|x\rangle \approx \frac{\sum_{n=1}^{N} t^n \exp\left[-\frac{\left(\theta^n\right)^2}{2\sigma_g^2}\right]}{\sum_{n=1}^{N} \exp\left[-\frac{\left(\theta^n\right)^2}{2\sigma_g^2}\right]} \tag{27}$$

The equation expresses that, for an input pattern $\mathbf{x}$, the output provided by the model can be viewed as a weighted average of all the target values contained in the training set. The weight for t^n is computed according to the Euclidean distance of the input to its corresponding pattern $\mathbf{x}^n$.

The variance σ_g acts as a smoothing parameter and its value must be specified by the designer. Parzen (1962) and Cacoullos (1966) showed that the estimator used for $p(\mathbf{x}, t)$ is consistent, i.e., asymptotically converges to the underlying probability density function $p(\mathbf{x}, t)$, at all points $(\mathbf{x}, t)$ at which the density function is continuous, provided that $\sigma_g = \sigma_g(N)$ is chosen as a decreasing function of N. The variance must meet the following requirements:

$$\lim_{N\to\infty} \sigma_g(N) = 0 \tag{28}$$

and

$$\lim_{N\to\infty} N\sigma_g^d(N) = \infty \tag{29}$$

When the parameter σ_g is large, the estimated density is forced to be smooth and, in the limit, becomes a multivariate Gaussian with covariance matrix $\sigma_g = I$. In this case, the output of the model tends to be the mean value of the target samples in the training set. As σ_g goes to

0, the output assumes the value of the target t^n associated with the observation $\mathbf{x}^n$ closest to the input $\mathbf{x}$. For intermediate values of σ_g, all values of t^n are taken into account, but those corresponding to points closer to $\mathbf{x}$ are given heavier weights.

4. RESULTS

4.1. Performance Assessment

The intraclass correlation coefficient (ICC) was used to evaluate the performance of the GRNN-based algorithms for AHI estimation. ICC is a reliability index used to evaluate the agreement between quantitative scores from different observers (Fisher, 1925). Unlike the Pearson correlation coefficient, ICC takes systematic error into account in assessing agreement.

The model (2,1) described by Shrout and Fleiss (1979) was used to compute the ICC. The obtained value can be interpreted as the proportion of relevant variance that is associated with differences among measured subjects and ranges from −1 to 1. A negative ICC indicates that more differences are observed within (error in the approximation) than between subjects. ICC values close to one reflect good reliability. However, it is worth noting that its value depends on the variability between subjects. Low levels of between-subjects variability will reduce the ICC even if the differences between subjects' scores across observers are small (Weir, 2005). A detailed description on the computation of ICC can be found in the study by Shrout and Fleiss (1979).

4.2. Training Set

As described previously, GRNN algorithms are based on the estimation of the density $p(x, i)$ using the Parzen window technique, which requires the definition of the functional form of the kernel. In our study, Gaussian kernels were used since they prevent discontinuities due to their smooth profile (Bishop, 1995). The bandwidth parameter (σ_g), which is common for all the kernel functions, must be defined to implement these algorithms. It determines the complexity of the mapping and, thus, has strong effects on the generalization ability of the algorithm. The goal is to find models that give good predictions for new data. The bias-variance trade-off shows that there is an optimum degree of model complexity for a given training set. Models with too little flexibility as well as those with a high degree of complexity will lead to poor generalization ability due to underfitting and overfitting, respectively (Bishop, 1995). Thus, a model selection process must be carried out to properly design pattern recognition algorithms, i.e. to find the optimum balance between bias and variance.

Model selection requires to compare the performance of several GRNN algorithms with different configurations (i.e. with a different value of the design parameter σ_g). Model selection was performed by applying leave-one-out cross-validation from training data (Jain, 2000). For a given configuration, a GRNN model was implemented by using $N - 1$ training pairs from the original set D. This model was then used to estimate the AHI for the excluded feature pattern. This process is repeated N times for all the samples in the training set. As a result, a prediction is obtained for each of the SaO_2 signals in the training set. The predicted values can be used to estimate the performance of the specified configuration. Finally, the optimum σ_g is selected by comparing the performance of different model configurations through the leave-one-out procedure.

The spread value (Δ), defined as the distance from the centre for which the output of the Gaussian kernel is 0.5, was used instead of σ_g to analyze the effect of the bandwidth on generalization capability. A Gaussian kernel centered on each training sample was used to compute the approximated density. The mean distance (r^n) of each center to its K-nearest centers (K = 3) was computed. The average (r_{avg}) for the set of r^n values was used to determine Δ. A proportional constant τ was defined as indicated in the following equation:

$$\Delta = \tau r_{avg} \tag{30}$$

The kernel bandwidth has a direct influence on the performance of the regression algorithm. As σ_g is increased, the approximated density becomes smoother and tends to a Gaussian profile.

Conversely, it is more irregular as σ_g is smaller. A wide range of σ_g values were evaluated by varying τ from 0.1 to 5. Figure 3 shows ICC as a function of the kernel bandwidth parameter. The highest ICC was achieved with τ = 0.5. Thus, this value was selected as the optimum bandwidth parameter for the AHI estimation problem.

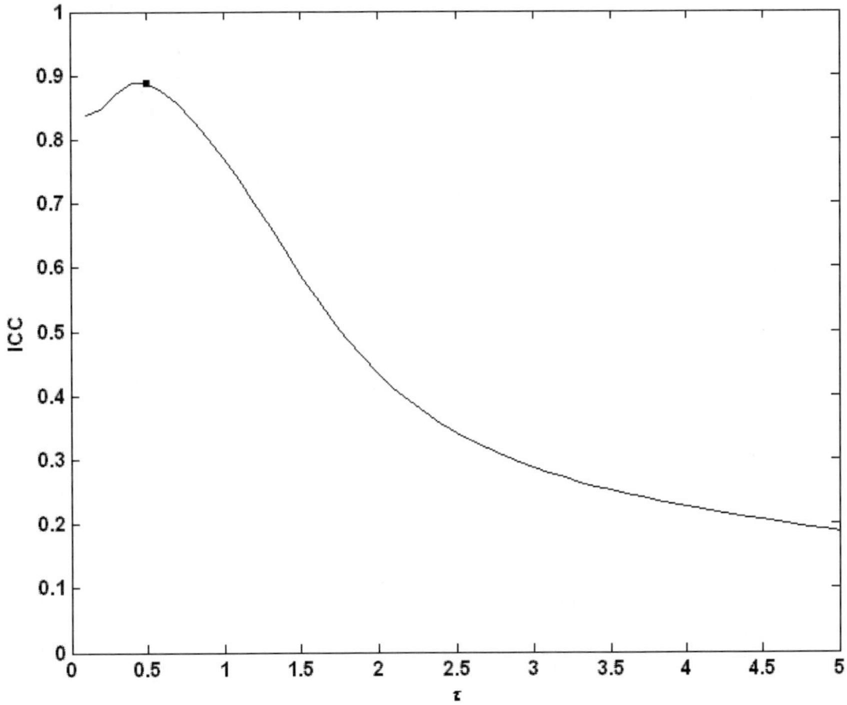

Figure 3. Influence of the spread parameter (controlled by τ) on the *ICC*.

4.3. Test Set

Once model selection was completed, a new model with the selected σ_g was trained by using the whole training set. The implemented GRNN model was evaluated on the test set with 144 SaO_2 signals, yielding an ICC of 0.89. Additionally, the utility of the estimated AHI to identify SAHS severity was evaluated. According to the estimation provided by the algorithm, each subject in the test set was assigned to one of the four following groups: normal ($0 \leq AHI < 5\ h^{-1}$), mild SAHS ($5 \leq AHI < 15\ h^{-1}$), moderate SAHS ($15 \leq AHI \leq 30\ h^{-1}$) and severe SAHS ($AHI > 30\ h^{-1}$). The confusion matrix obtained for the proposed GRNN model is provided in Table 2. As can be observed, the GRNN-based algorithm showed high reliability for identifying subjects with mild, moderate and severe SAHS.

Table 2. Capability of the GRNN-based estimation algorithm to assess SAHS severity. Confusion matrix for classification of SaO_2 signals into four different groups: normal ($0 \leq AHI < 5\ h^{-1}$), mild SAHS ($5 \leq AHI < 15\ h^{-1}$), moderate SAHS ($15 \leq AHI \leq 30\ h^{-1}$) and severe SAHS ($AHI > 30\ h^{-1}$)

GRNN				
True	Predicted			
	No SAHS	Mild	Moderate	Severe
No SAHS	6	23	5	0
Mild	1	23	8	0
Moderate	0	6	20	4
Severe	0	0	6	42

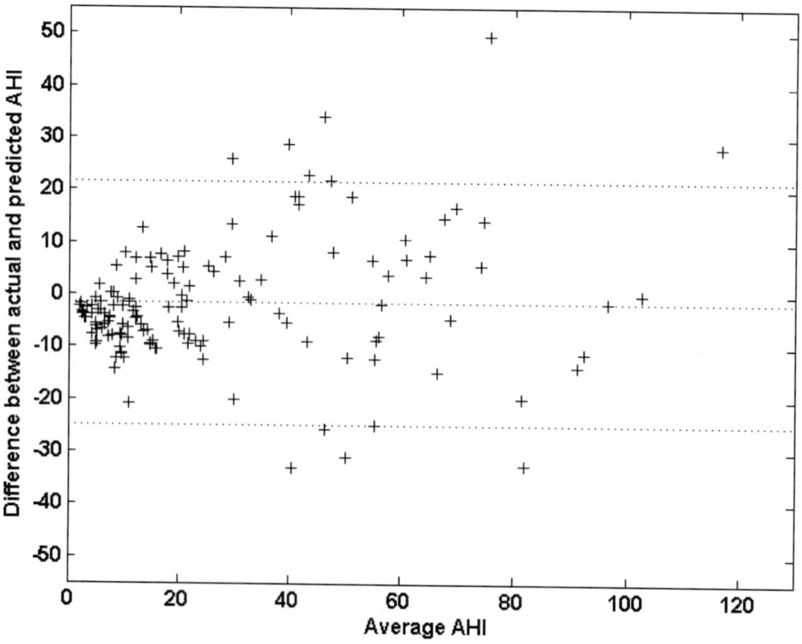

Figure 4. Bland-Altman plot from the AHI estimations provided by the GRNN algorithm on signals in the test set.

However, our results reveal that it tends to label normal subjects as mild-SAHS. In addition, Bland-Altman analysis was carried out for further assessment of the proposed AHI estimation method. The difference between actual AHI and predicted AHI for each subject is depicted as a function of the mean value of both quantities. The mean and the limits of the 95% confidence interval for the difference provide useful information about the quality of the approximation. The obtained Bland-Altman plot is shown in Figure 4. As reflected by the plot, the average difference (bias) between both scores of AHI is close to zero. Although a negative value of the bias was obtained, its reduced magnitude indicates that there was not any marked trend reflecting overestimation effects. On the other hand, it should be noted that the plot shows a trend in the variability of the difference since the scatter gets larger as the average gets higher. This behavior reflects that the error becomes larger for higher values of AHI, what could be expected in the analyzed problem.

5. DISCUSSION

A pattern recognition-based solution has been proposed for automatic assessment of from SaO_2 data recorded through nocturnal pulse oximetry. A regression approach was adopted to model SAHS diagnosis. A methodology composed of three different stages was designed to estimate the AHI from SaO_2 analysis: signal preprocessing, feature extraction and regression analysis. Both time-domain and frequency-domain features were defined to characterize SaO_2 signals. Subsequently, a GRNN model was trained to estimate the AHI from the set of extracted features. The performance of the proposed estimation algorithm was evaluated using the ICC and Bland-Altman plots. Furthermore, a confusion matrix was obtained in order to assess the utility of the algorithm to predict SAHS severity.

Our algorithm for AHI estimation yielded promising results in the SAHS diagnosis problem. An ICC of 0.89 was achieved, reflecting high reliability in the estimated AHI. Previously, diverse methods for automatic interpretation of SaO_2 signals in the context of SAHS diagnosis have been proposed. However, most of them adopted a two-class classification approach based on labeling subjects as SAHS-negative or SAHS-positive. Oxygen desaturation indices represent a straightforward technique for this purpose. Indeed, these indices are commonly used in clinical practice since some commercial oximeters include the software to compute them (McNicholas, 2008). Gyulay et al. (1993) assessed the diagnostic utility of ODI3, ODI4 and CT90 on a population of 98 subjects. Marked differences between sensitivity and specificity statistics were found for each of them. The highest sensitivity was achieved by CT90 with a value of 93% whereas its specificity was 51%. ODI4 provided a specificity of 98%. In contrast, a poor sensitivity value of 40% was reached. More balanced diagnostic results of this parameter were reported by other researchers. A sensitivity of 71% and a specificity of 93% were observed by Chiner et al. (1999) using data from 275 subjects. Oximetry signals from a larger population (359 subjects) were available for the study of Gurubhagavatula et al. (2001). ODI4 reached 85.2% sensitivity and 90% specificity when the criterion to define SAHS was AHI $\geq 5\ h^{-1}$. Other methods based on statistical data analysis and signal processing have been also applied to SaO_2 recordings for identifying SAHS-negative and SAHS-positive subjects. Zamarrón et al. (1999) analyzed the properties of SaO_2 signals in the frequency domain. They found that the spectrum of recordings corresponding to SAHS-positive subjects tends to have a peak between 0.010 and 0.033 Hz. This result was used to define some of the frequency-domain features proposed in the present study. A similar behavior was subsequently reported for HRV signals (Zamarrón et al. 2003). The dynamical properties of SaO_2 signals were also evaluated with non-linear methods. It was found that recordings corresponding to SAHS-positive subjects presented higher irregularity, variability and complexity using ApEn, CTM and LZC, respectively (Álvarez et al. 2006; Del Campo et al. 2006; Hornero et al. 2007). Indeed, the results reported by these studies motivated the use of these non-linear methods to characterize oximetry data in this study. However, the utility of both spectral and non-linear features derived from oximetry recordings to predict SAHS severity was not evaluated in these previous studies.

Although two-class classification approaches as those previously cited provide relevant results in the SAHS diagnosis problem, more valuable information could be derived from a regression model for AHI estimation. The regression approach adds useful knowledge about the degree of severity of SAHS. Additionally, the estimated AHI can be extended to different cut-off points to define SAHS (there is no consensus about this issue and the threshold used in the literature varies from 5 h^{-1} to 15 h^{-1}) (Caples et al. 2005). Conventionally, subjects referred to sleep study are assigned to one of the four following categories: normal, mild, moderate and severe SAHS. A confusion matrix has been built in order to evaluate the utility of the AHI estimated by our GRNN-based model to identify subjects from these four groups. It was noticed that the algorithm accurately recognized subjects with mild, moderate or severe SAHS. Nevertheless, the main weakness of our method was the classification of normal cases. As reflected by our results, the GRNN estimator tends to label normal subjects as mild-SAHS. This situation was observed for 23 out of 34 normal subjects in the test set. Despite this undesired effect, the proposed method reflects a significant capability as a screening tool due to the high accuracy reached for the remaining three categories. In SAHS screening, false negative cases must be avoided in order to prevent subjects from other health complications because of the lack of treatment. It is worth noting that only one subject with mild SAHS and none of the subjects with moderate or severe SAHS were identified as normal. This is an essential result in order to avoid false negative diagnosis.

A reduced number of studies focused on the estimation of the AHI have been proposed. Magalang et al. (2003) developed a predictor model for AHI based on conventional oximetry indices and multivariate adaptive regression splines (MARS). It achieved high correlation (0.84) with a sensitivity of 90% and a specificity of 70% in SAHS detection. Other signals and data different to oximetry have been analyzed for this purpose. Kirby et al. (1999) developed a regression model to estimate the AHI from clinical features using GRNN. Sensitivity and specificity were 98.9% and 80%, respectively. However, results about the quality of the estimations were not provided. Another regression algorithm based on neural networks was developed by El-Solh et al. (1999). Clinical and anthropomorphic features were used together with a MLP network. This regression algorithm achieved a correlation of 0.86 with the AHI derived from PSG. It outperformed linear approximation using MLR. As can be observed, our GRNN-based model outperformed previous methods for AHI estimation. Thus, it can be viewed as a robust technique for automatic computation of the AHI only using oximetry data.

Some limitations can be pointed out in our study. The size of the database can be considered as a constraint. GRNN algorithms represent highly complex (flexible) pattern recognition methods and could be affected by overfitting. It is required to control model complexity to prevent this phenomenon. An exhaustive model selection search was performed in order to find the optimum value of σ_g, which determines the complexity of the GRNN algorithm for a given training set. The presence of other respiratory disorders different to SAHS may lead to an incorrect estimation of the AHI. For instance, chronic obstructive pulmonary disease (COPD) can affect the dynamical behavior of oximetry recordings similarly to SAHS. Finally, SaO_2 signals analyzed in this study were recorded during in-hospital evaluation of the patients. In order to assess the utility of ambulatory pulse oximetry, SaO_2 signals recorded at home should be considered.

CONCLUSION

In summary, SAHS diagnosis has been modeled as a regression task. An algorithm for automatic AHI estimation from oximetry data has been developed using GRNN. The implemented method represents an efficient novel tool to help in SAHS detection since high reliability was achieved (ICC = 0.89). The proposed regression approach provides information about the degree of severity of SAHS instead of simply labeling subjects as SAHS-negative or SAHS-positive. It represents an added value when compared with two-class classification algorithms for SAHS diagnosis. Therefore, the proposed GRNN-based method could be considered as a screening tool for SAHS in order to reduce the number of required PSG studies.

ACKNOWLEDGMENTS

This research was supported in part by project VA111A11-2 from Consejería de Educación (Junta de Castilla y León), Ministerio de Economía y Competitividad and FEDER under project TEC2011-22987 and the Proyecto Cero 2011 on Ageing from Fundación General CSIC, Obra Social La Caixa and CSIC.

REFERENCES

Álvarez, D., Hornero, R., Abásolo, D., Del Campo, F. and Zamarrón, C. (2006) Nonlinear characteristics of blood oxygen saturation from nocturnal oximetry for obstructive sleep apnoea detection. *Physiological Measurement*, 27, 399–412.

Bennet, J.A. and Kinnear, W.J.M. (1999) Sleep on the cheap: the role of overnight oximetry in the diagnosis of sleep apnoea hypopnoea syndrome. *Thorax*, 11, pp. 958–959.

Bishop, C.M. (1995) *Neural Networks for pattern recognition*. Oxford: Oxford University Press.

Bloch, K.E. (2003) Getting the most out of nocturnal pulse oximetry. *Chest*, 124, pp. 1628–1630.

Cacoullos, T. (1966) Estimation of a multivariate density. *Annals of the Institute of Statistical Mathematics*, 18, pp. 179–189.

Caples, S.M., Gami, AS and Somers, V.K. (2005) Obstructive sleep apnea. *Annals of Internal Medicine*, 142, pp. 187–197.

Chiner, E., Signes-Costa, J., Arriero, J.M., Marco, J., Fuentes, I. and Sergado, A. (1999) Nocturnal oximetry for the diagnosis of the sleep apnoea hypopnoea syndrome: a method to reduce the number of polysomnographies?. *Thorax*, 54, pp. 968–971.

Cohen, M.E., Hudson, D.L. and Deedwania, P.C. (1996) Applying continuous chaotic modelling to cardiac signals. *IEEE Engineering in Medicine and Biology Magazine*, 15, pp. 97–102.

Del Campo, F., Hornero, R., Zamarrón, C., Abásolo, D. and Álvarez, D. (2006) Oxygen saturation regularity analysis in the diagnosis of obstructive sleep apnea. *Artificial Intelligence in Medicine*, 37, 111–118.

El-Solh, A.A., Mador, M.J., Ten-Brock, E., Shucard, D.W., Abul-Khoudoud, M. and Grant, B.J.B. (1999) Validity of neural network in sleep apnea. *Sleep*, 22, pp. 105–111.

Firat, M., Tran, M.E. and Yurdusev, M.A. (2010) Comparative analysis of neural network techniques for predicting water consumption time series Mahmut Firat, Mustafa Erkan Turan, Mehmet Ali Yurdusev, *Journal of Hydrology*, 384, pp. 46–51.

Fisher, R.A. (1925) *Statistical methods for research workers*. Edinburgh: Oliver and Boyd.

Flemons, W.W., Douglas, N.J., Kuna, S.T., Rodenstein, D.O. and Wheatley, J. (2004) Access to diagnosis and treatment of patients with sleep apnea. *Amercian Journal of Respiratory and Critical Care Medicine*, vol. 169, pp. 668–672.

Gagnadoux, F., Pelletier-Fleury, N., Philippe, C., Rakatonanahary, D. and Bernard, F. (2002) Home unattended vs hospital telemonitored polysomnography in suspected obstructive sleep apnea syndrome: a randomized crossover trial. *Chest*, 121, pp. 753–758.

Gurubhagavatula, I., Maislin, G. and Pack, A.I. (2001) An algorithm to stratify sleep apnea risk in a sleep disorders clinic population. *American Journal of Respiratory and Critical Care Medicine*, 164, pp. 1904–1909.

Gyulay, S., Olson, L.G., Hensley, M.J., King, M.T., Allen, K.M. and Saunders, N.A. (1993) A comparison of clinical assessment and home oximetry in the diagnosis of obstructive sleep apnea. *The American Review of Respiratory Disease*, 147, pp. 50–53.

Hornero, R., Álvarez, D., Abásolo, D., Del Campo, F. and Zamarrón, C. (2007) Utility of approximate entropy from overnight pulse oximetry data in the diagnosis of the obstructive sleep apnea syndrome,' *IEEE Transactions on Biomedical Engineering*, 54, pp. 107–113.

Jain, A.K., Duin, R.P.W. and Mao, J. (2000) Statistical pattern recognition: a review. *IEEE Transactions on Pattern Analysis and Machine Intelligence*, 22, pp. 4–37.

Jobson, J.D. (1991) *Applied multivariate data analysis. Volume I: regression and experimental design*. New York: Springer.

Kirby, S.D., Danter, W., George, C.F.P., Francovic, T., Ruby, R.P.F. and Ferguson, K.A. (1999) Neural network prediction of obstructive sleep apnea from clinical criteria. *Chest*, 116, pp. 409–415.

Kushida, C.A., Littner, M.R., Morgenthaler, T., Alessi, C.A., Bailey, D., Coleman, J., Friedman, L., Hirshkowitz, M., Kapen, S., Kramer, M., Lee-Chiong, T., Loube, D.L., Owens, J., Pancer, J.P. and Wise, M. (2005) Practice parameters for the indications for polysomnography and related procedures: an update for 2005. *Sleep*, 48, pp. 499–521.

Lee, W.Y., House, J.M. and Kyong, N.H. (2004) Subsystem level fault diagnosis of a building's air-handling unit using general regression neural networks, *Applied Energy*, 77, pp. 153–170.

Lempel, A. and Ziv, J. (1976) On the complexity of finite sequences. *IEEE Transactions on Information Theory*, 22, pp. 75–81.

Magalang, U.J., Dmochowski, J., Veeramachaneni, S., Draw, A., Mador, M.J., El-Solh, A. and Grant, B.J.B. (2003) Prediction of the apnea-hypopnea index from overnight pulse oximetry. *Chest*, 124, pp. 1694–1701.

Marcos, J.V., Hornero, R., Álvarez, D., Del Campo, F., López, M. and Zamarrón, C. (2008a) Radial basis function classifiers to help in the diagnosis of the obstructive sleep apnoea syndrome from nocturnal oximetry. *Medical and Biological Engineering and Computing*, 46, pp. 323–332.

Marcos, J.V., Hornero, R., Álvarez, D., Del Campo, F., Zamarrón, C. and López, M. (2008b) Utility of multilayer perceptron neural network classifiers in the diagnosis of the obstructive sleep apnoea syndrome from nocturnal oximetry. *Computer Methods and Programs in Biomedicine*, 72, pp. 79–89.

Marcos, J.V., Hornero, R., Álvarez, D., Del Campo, F. and Zamarrón, C. (2009) Assessment of four statistical pattern recognition techniques to assist in obstructive sleep apnoea diagnosis from nocturnal oximetry. *Medical Engineering and Physics*, 31, pp. 971–978.

Marcos, J.V., Hornero, R., Álvarez, D., Nabney, I.T., Del Campo, F. and Zamarrón, C. (2010a) The classification of oximetry signals using Bayesian neural networks to assist in the detection of obstructive sleep apnoea syndrome. *Physiological Measurement*, 31, pp. 375–394.

Marcos, J.V., Hornero, R., Álvarez, D., Del Campo, F. and Aboy, M. (2010b) Automated detection of obstructive sleep apnoea syndrome from oxygen saturation recordings using linear discriminant analysis. *Medical and Biological Engineering and Computing*, 48, pp. 895–902.

Marcos, J.V., Hornero, R., Álvarez, D., Aboy, M. and Del Campo, F. (2012) Automated prediction of the apnea-hypopnea index from nocturnal oximetry recordings, *IEEE Transactions on Biomedical Engineering*, 59, pp. 141-149.

McNicholas, W.T. (2008) Diagnosis of obstructive sleep apnea in adults. *Proceedings of the American Thoracic Society*, 5, pp. 154–160.

Moyle, J.T. (1996) Uses and abuses of pulse oximetry. *Archives of Disease in Childhood*, 74, pp. 77–80.

Netzer, N., Eliasson, A.H., Netzer, C. and Kristo, D.A. (2001) Overnight pulse oximetry for sleep-disordered breathing in adults: a review. *Chest*, 120, pp. 625–633.

Parzen, E. (1962) On estimation of a probability density function and mode. *Annals of Mathematical Statistics*, 33, pp. 1065–1076.

Philips Respironics (2010) Alice 5 Diagnostic Sleep System, Philips Respironics [WWW] Available from: http://alice5.respironics.com/ [Accessed 15/12/2010].

Pincus, S.M. (2001) Assessing serial irregularity and its implications for health. *Annals of the New York Academy of Science*, 954, pp. 245–267.

Qureshi, A. and Ballard, R.D. (2003) Obstructive sleep apnea. *The Journal of Allergy and Clinical Immunology*, 112, pp. 643–651.

Rechtschaffen, A and Kales, A (1968) *A manual of standardized terminology, techniques and scoring system for sleep stages of human subjects*. Los Angeles: Brain Information Services, Brain Research Institute, University of California.

Shrout, P.E. and Fleiss, J.L. (1979) Intraclass correlations: uses in assessing rater reliability. *Psychological Bulletin*, 36, pp. 420–428.

Specht, D.F. (1991) A general regression neural network. *IEEE Transactions on Neural Networks*, 2, pp. 568–576.

Weir, J.P. (2005) Quantifying test-retest reliability using the intraclass correlation coefficient and the SEM. *Journal of Strength and Conditioning Research*, 19, pp. 231–240.

Welch, P. (1967) The use of fast Fourier transform for the estimation of power spectra: a method based on time averaging over short, modified periodograms. *IEEE Transactions on Audio and Electroacoustics*, 15, pp. 70–73.

Yan, W. (2012) Toward automatic time-series forecasting using neural networks, *IEEE Transactions on Neural Networks and Learning Systems*, 23, pp. 1028-1039.

Young T., Evans L., Finn L. and Palta M (1997) Estimation of the clinically diagnosed proportion of sleep apnea syndrome in middle-aged men and women. *Sleep*, 20, pp. 705–706.

Zamarrón, C., Romero, P.V., Rodríguez, J.R. and Gude, F. (1999) Oximetry spectral analysis in the diagnosis of obstructive sleep apnoea. *Clinical Science*, 97, pp. 467–473.

Zamarrón, C., Gude, F., Barcala, J., Rodríguez, J.R. and Romero, P.V. (2003) Utility of oxygen saturation and heart rate spectral analysis obtained from pulse oximetric recordings in the diagnosis of sleep apnea syndrome. *Chest*, 123, pp 1567–1576.

In: Focus on Nonlinear Analysis Research
Editors: Gualtiero Padovani and Manuel Occhino
ISBN: 978-1-62417-350-9
© 2013 Nova Science Publishers, Inc.

Chapter 3

NONLINEARITY IN BRIDGE STRUCTURAL ANALYSIS

Chung C. Fu* and Mohamed S. Ahmed
University of Maryland at College Park, MD, US

ABSTRACT

The term "nonlinear system" refers to a given structure undergoing a change in stiffness in its loaded state. For linear analyses, intermediate steps between stages are not required. However, this is a necessity when dealing with nonlinear problems.

There are, in general, two types of nonlinearity. The first type is called "material nonlinearity", in which the nature of the material changes the stiffness of the structure as it is loaded. The second type is called "geometric nonlinearity", in which a given structure undergoes large displacements relative to the size of the structure. This causes large changes in the geometric stiffness of the structure. Material nonlinearity covers the subjects of the stress-strain relationships corresponding to unconfined concrete, confined concrete, and longitudinal steel reinforcement and others, such as Bauschinger effect, stress relaxation and creep. Geometric nonlinearity includes second-order effects on a structure, as well as instability of a large structure. The original deflection theory and modern matrix analyses with geometric nonlinearities for long-span bridges are discussed, respectively. Other types of nonlinearity such as nonlinearity due to change of status are also covered. In order to demonstrate the usage, nonlinear analyses of long-span bridges (cable stayed bridges and suspension bridges) and nonlinear seismic analyses of bridge structures (nonlinear time history analysis and modal pushover analysis with their comparison) are also given in this chapter.

Nowadays, with the rapid development of modern day software, many sophisticated structural programs can be used for solving nonlinear problems under extreme limit states to complete the bridge design and checking process. With these tools, nonlinear problem solving becomes more or less a routine process for structural engineers. Through this article, engineers may become familiar with those complexities and assumptions to analyze bridges in the nonlinear range to make them safe and economical.

---

* Corresponding author (ccfu@umd.edu).

1. INTRODUCTION

Generally, we may treat a bridge structure as a linear system, which means its output (stresses, strains, deformations, forces, and displacements of bridge components) is directly proportional to its input (loads and settlement). However, under certain circumstances, we may encounter problems when the bridge structure does not behave elastically.

The linearity is due to the fact that the stiffness of the material does not change in a loaded state. In general, all metals behave in a linear fashion within a certain stress range. On the other hand, the term "nonlinear system" refers to a given structure undergoing a change in stiffness in its loaded state. Nonlinearity in structural terms is the system that does not satisfy the superposition principle in physics. There are, in general, two types of nonlinearity. The first type is called "material nonlinearity", in which the nature of the material changes the stiffness of the structure as it is loaded. The second type is called "geometric nonlinearity", in which a given structure undergoes large displacements relative to the size of the structure. This causes large changes in the geometric stiffness of the structure (AASHTO 2012).

Material nonlinearity includes inelastic behavior of elements and cross sections due to nonlinear material stress-strain relationship or nonlinear gaps, dampers, and springs in special bridge components. For steel elements, yielding is the cause of nonlinearity and should be taken into account in the analysis. Strain hardening and fracturing of steel may also be considered. For concrete elements, nonlinear strain–stress relationship in compression up to the ultimate strain and beyond, ignoring the concrete tension strength, may be considered and simulated. Geometric nonlinearity includes second-order effects on a structure, as well as instability of a large structure. A first-order, or linear, analysis is usually adequate for short- to medium-span bridges, which consist 90% or more of the bridge population. However, a second-order, or nonlinear analysis is recommended for long-span, tall, and slender bridges. A truly large displacement analysis is generally required for suspension bridges.

Moreover, with the advance of the seismic analysis, linear analysis with strength-based design may not be enough for high seismic hazard areas; nonlinear analyses with displacement-based design should be used. Other nonlinearities, such as soil–structural interaction (SSI), seismic response modification devices (dampers and seismic isolations), connection flexibility, gap close and opening should be carefully considered in the analysis.

There are key steps and advices when modeling a structure using nonlinear analysis: understanding the physical problem and concepts that underlie the analysis procedure, understanding the software capability and limitations, and critical examination of the analysis output, etc. When invoking a nonlinear analysis model, careful consideration and prudent judgment should be used before implementation. The utilization of the superposition principle is not applicable to nonlinear problems. When the load is doubled, this produces more, or less, than double the response. In relation to this principle, the results of different applied load cases cannot be added together. If the different loads are applied in a sequential manner, the end result of the stress state and deformation could depend on the order and manner in which the loads were applied to the structure. A linear model should be performed prior to a nonlinear model. Equally important, load steps are critical to nonlinear analysis. The final load must be approached in several steps. If a load step is sufficiently large, this can slow the convergence of the model as well as provide a drastic change in the load versus displacement plot. Each load step can produce a significant amount of results that should be reviewed and analyzed. Warnings and errors from the computer analysis should also be reviewed with careful attention.

2. Analytical Approach

Linearity and Nonlinearity Iterations

For linear problems, intermediate steps between stages are not required. However, this is a necessity when dealing with nonlinear problems. The "Newton-Raphson" approach is the most popular method used to solve nonlinear problems. In this approach, the load is subdivided into a series of load increments. The load increments can be applied over several load steps. Figure 1 illustrates the use of Newton-Raphson equilibrium iterations in a single degree-of-freedom nonlinear analysis. The procedure for analyzing nonlinear transient behavior is similar to that used for nonlinear static behavior; application of the load is done in incremental steps, and the program performs equilibrium iterations at each step (ANSYS 1992).

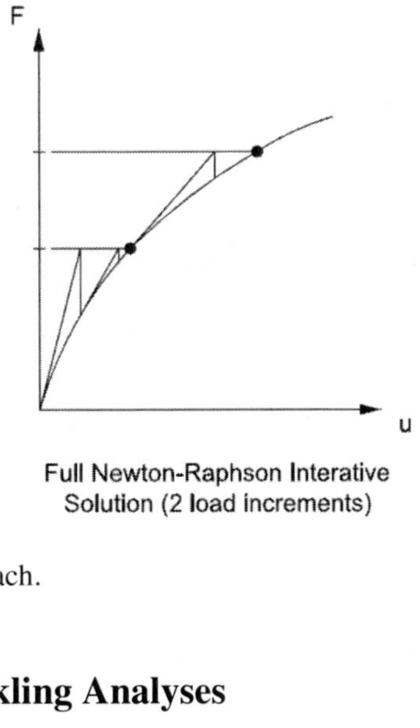

Figure 1. Newton-Raphson approach.

Elastic and Inelastic Buckling Analyses

Eigenvalue analysis predicts the theoretical buckling strength of a structure idealized as elastic. In the classic eigenvalue method, structural eigenvalues are computed from the loading conditions and constraints of a given system. For a basic structural configuration, tabulated solutions provide buckling loads. Each load has an associated buckled mode shape which represents the shape a structure assumes when buckled. During elastic, or linear, buckling analysis, perturbations are applied to the un-deformed configuration of a structure. A specified set of loads are observed for which deflections could induce instability under P-Delta effect. Elastic buckling analysis produces a set of buckling factors and corresponding mode shapes. The structure will buckle at a load equal to the applied load, in the linear eigenvalue prediction step, multiplied with the obtained eigenvalue. Similarly, the mode shapes are normalized displacement sets which indicate the configuration of the buckled structure (ANSYS 1992). However, in a real bridge structure, imperfections and nonlinear behavior keep the system from achieving this theoretical buckling strength, leading eigenvalue analysis to over-predict buckling load. Inelastic, or nonlinear, buckling analysis is therefore recommended.

Nonlinear buckling analysis provides greater accuracy by incrementally increasing load application until a structure becomes unstable. This condition of instability is achieved when a small increase in the load level causes a very large change in displacement. Nonlinear buckling analysis is a static method which accounts for material and geometric nonlinearities (P-Δ and P-δ effects), load perturbations, geometric imperfections, and gaps. Either a small destabilizing load or an initial imperfection is necessary to initiate the solution of a desired buckling mode.

A nonlinear analysis requires incremental load (or displacement) steps in an explicit or an implicit manner. At the end of each increment the structure geometry changes and possibly the material is nonlinear or the material has yielded. An explicit nonlinear analysis performs the incremental procedure and at the end of each increment updates the stiffness matrix based on geometry changes (if applicable) and material changes (if applicable). An implicit nonlinear analysis does the same thing but uses Newton-Raphson iterations to enforce equilibrium, which is the primary difference between the two types of analyses. Either explicit or implicit nonlinear static analysis can be used. However, for nonlinear buckling analysis, the implicit method is preferred.

3. Material Nonlinearity

Stress-Strain Relationships

Nonlinear stress-strain relationships of plastic, multi-linear elastic, and hyper-elastic materials will cause a structure's stiffness to change at different load levels (and, typically, at different temperatures). Creep, visco-plasticity, and viscoelasticity will give rise to nonlinearities that can be time-, rate-, temperature-, and stress-related. Most common engineering materials exhibit a linear stress-strain relationship up to a stress level known as the "proportional limit". Beyond this limit, the stress-strain relationship will become nonlinear (Figure 2). Plastic behavior, characterized by non-recoverable strain, begins when stresses exceed the material's yield point.

For concrete elements, the strain–stress relationship in compression is nonlinear up to the ultimate strain and beyond. Several models for the stress-strain relationship of concrete have been proposed in the past. At low levels of stress, transverse reinforcement is hardly stressed; the concrete behaves much like "unconfined" concrete. At stresses close to the uniaxial strength of concrete internal fracturing causes the concrete to dilate and bear out against the transverse reinforcement which then causes a confining action in the concrete. This phenomenon of "confined" concrete by suitable arrangement of transverse reinforcement causes a significant increase in the strength and ductility of concrete. The enhancement of strength and ductility by confining the concrete is an important aspect that needs to be considered in the design of structural concrete members, especially in areas prone to seismic activity, blast effects or vehicle crashes. The stress-strain relationships corresponding to unconfined concrete, confined concrete, and longitudinal steel reinforcement are discussed in the following sections.

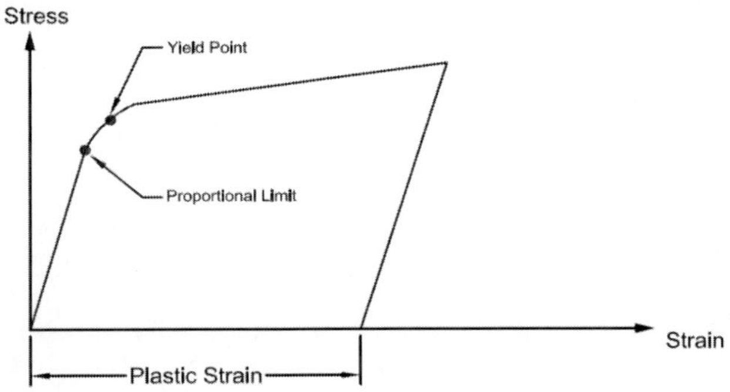

Figure 2. Elastic-inelastic stress-strain curve.

Unconfined and Confined Concrete

Numerous stress-strain relationships for unconfined and confined concrete were developed. The two most popular ones based on their usage are listed here. Kent and Park (1971) proposed a stress-strain equation for both unconfined and confined concrete. In their model, they generalized Hognestad's (1951) equation to describe in a more complete manner on the post-peak stress-strain behavior. In this model, the ascending branch is represented by modifying the Hognestad second degree parabola by replacing $0.85f_c'$ with f_c' and ε_{co} with 0.002. Kent and Park modified their model again in 1982 as shown in Figure 3.

$$f_c = f_c' \left[\frac{2\varepsilon_c}{\varepsilon_{co}} - \left(\frac{\varepsilon_c}{\varepsilon_{co}} \right)^2 \right] \qquad (1)$$

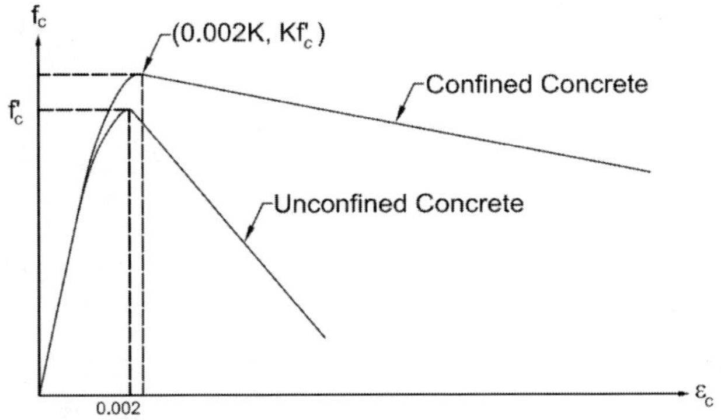

Figure 3. Stress-strain behavior of compressed concrete confined by rectangular steel hoops based on Modified Kent and Park model.

Mander et al. (1988a) first tested circular, rectangular and square full scale columns at seismic strain rates to investigate the influence of different transverse reinforcement arrangements on the confinement effectiveness and overall performance. Mander et al. (1988b) went on to model their experimental results. It was observed that if the peak strain and stress coordinates (ε_{cc}, f_{cc}') could be found, then the performance over the entire stress-strain range was consistent, regardless of the arrangement of the confinement reinforcement used. The equations are listed below and shown in Figure 4.

$$\frac{f_c}{f_{cc}'} = \frac{n\left(\frac{\varepsilon_c}{\varepsilon_{cc}}\right)}{(n-1)+\left(\frac{\varepsilon_c}{\varepsilon_{cc}}\right)^n} \tag{2}$$

in which

$$n = \frac{E_c}{E_c - E_{sec}} \tag{2a}$$

$$E_c(MPa) = 5000\sqrt{f_c'(MPa)} \tag{2b}$$

$$E_{sec} = \frac{f_{cc}'}{\varepsilon_{cc}} \tag{2c}$$

where ε_{cc} is the strain at the maximum compressive strength of confined concrete f_{cc}'

$$\varepsilon_{cc} = \varepsilon_{co}\left[1 + 5\left(\frac{f_{cc}'}{f_c'} - 1\right)\right] \tag{2d}$$

and f_{cc}', the compressive strength of confined concrete, is given as

$$f_{cc}' = f_c'\left(-1.254 + 2.254\sqrt{1 + \frac{7.94 f_t'}{f_c'}} - 2\frac{f_t'}{f_c'}\right) \tag{2e}$$

in which f_t' is given by

$$f_t' = \frac{1}{2} k_e \rho_s f_{yh} \tag{2f}$$

in which ρ_s= ratio of volume of transverse confining steel to volume of confined concrete core, f_{yh}= yield strength of transverse reinforcement, k_e = confinement coefficient.

For circular hoops

$$k_e = \frac{\left(1 - \frac{s'}{2d_s}\right)^2}{1 - \rho_{cc}} \tag{2g}$$

For circular spirals

$$k_e = \frac{1-\frac{s'}{2d_s}}{1-\rho_{cc}} \tag{2h}$$

where ρ_{cc}= ratio of area of longitudinal reinforcement to area of core of the section. s'= clear spacing between spiral of hoop bars, d_s = diameter of spiral.

Due to its generality, the Mander et al. (1988b) model (Figure 4) has enjoyed widespread use in design and research despite a few shortcomings.

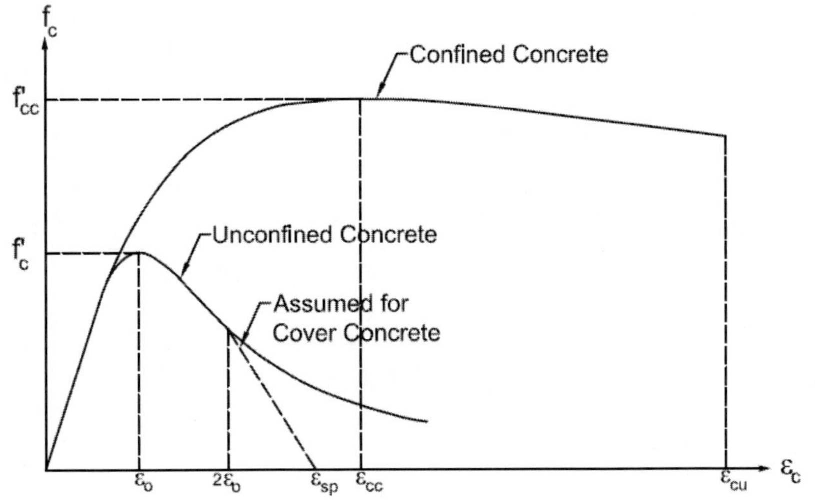

Figure 4. Stress-strain relation for monotonic loading of confined and unconfined concrete based on Mander et al. model.

Reinforcing Steel and Structural Steel

In the analysis of moments and axial loads, two different models of the stress-strain performance of the reinforcing steel may be adopted. For nominal design capacities, an elasto-plastic model is customarily adopted to provide a dependable estimate for design.

For "exact" analysis of existing reinforced concrete members, a realistic stress-strain model should be applied using expected values of the control parameters. Such a model (Figure 5), conveniently posed in the form of a single equation, is given as:

$$f_s = \frac{E_s\varepsilon_s}{\left\{1+\left|\frac{E_s\varepsilon_s}{f_y}\right|^{20}\right\}^{0.05}} + \left(f_{su}-f_y\right)\left[1-\frac{|\varepsilon_{su}-\varepsilon_s|^P}{\{|\varepsilon_{su}-\varepsilon_{sh}|^{20P}+|\varepsilon_{su}-\varepsilon_s|^{20P}\}^{0.05}}\right] \tag{3}$$

where

$$P = \frac{E_{sh}(\varepsilon_{su} - \varepsilon_{sh})}{(f_{su} - f_y)} \quad (3a)$$

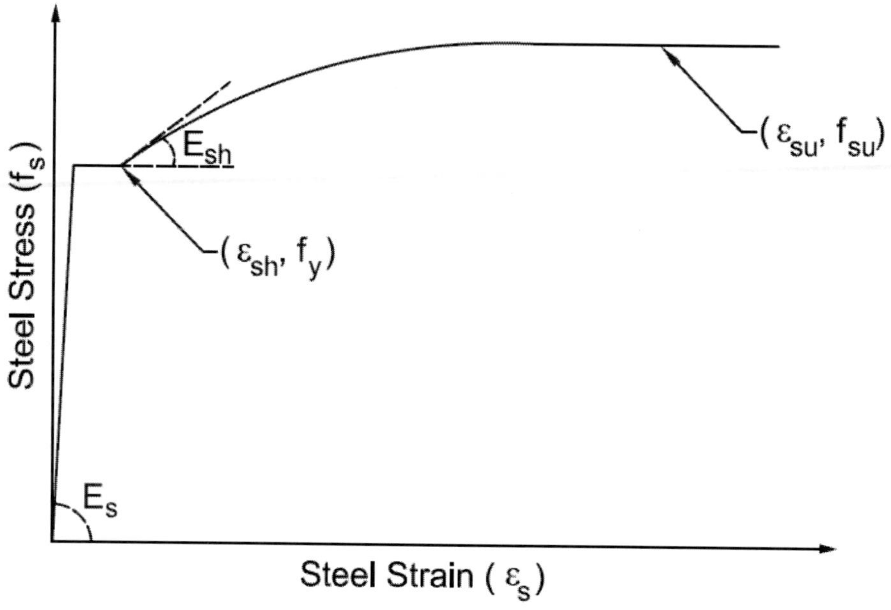

Figure 5. Stress-strain curve for steel.

For the longitudinal steel, a bilinear stress-strain relationship was estimated and employed (Figure 6).

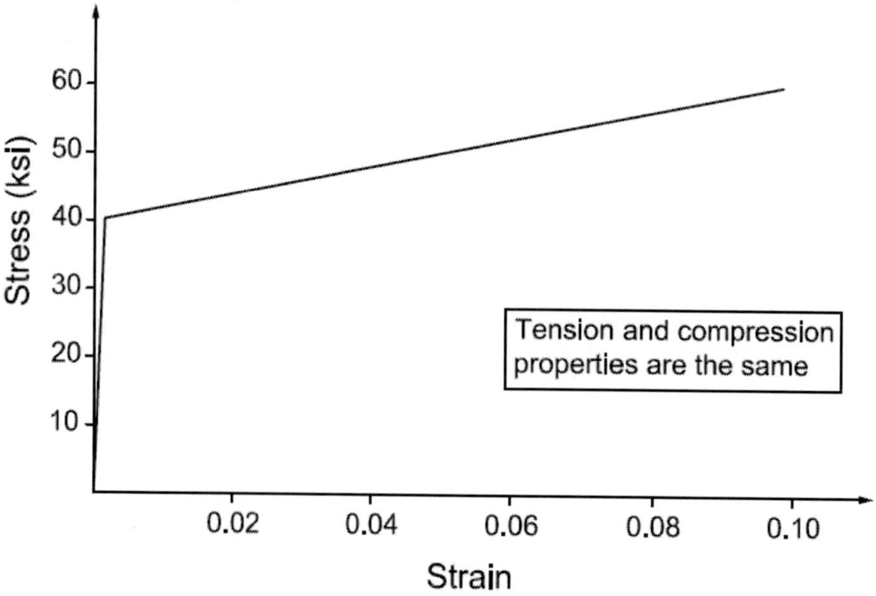

Figure 6. Idealized steel stress-strain relationships.

Others

The Multi-linear Isotropic Hardening (MISO) defines the Von Mises yield criteria coupled with an isotropic work hardening assumption. The Bilinear Isotropic Hardening is like the multi-linear isotropic hardening option, except that a bilinear curve is used instead of a multi-linear curve (Figure 7).

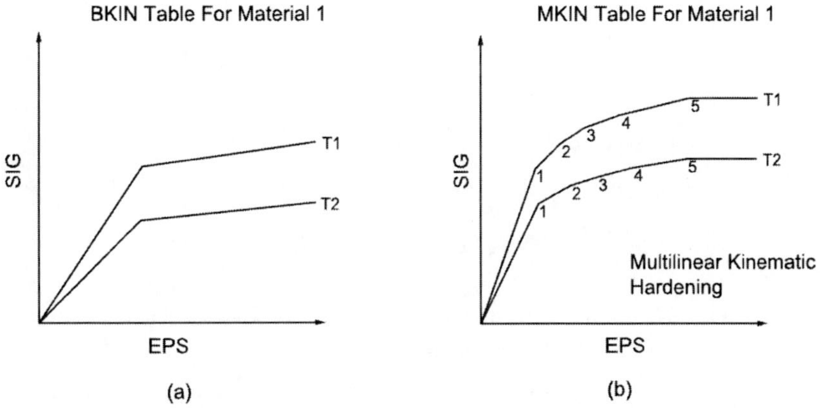

Figure 7. (a) Bilinear kinematic hardening (b) Multi-linear kinematic hardening.

Under cyclic loading, the Bauschinger effect, an increase in tensile yield strength with deceased compressive yield strength, may occur (see Figure 8). This phenomenon may apply for general small-strain use for materials that obey Von Mises yield criteria (which includes most metals).

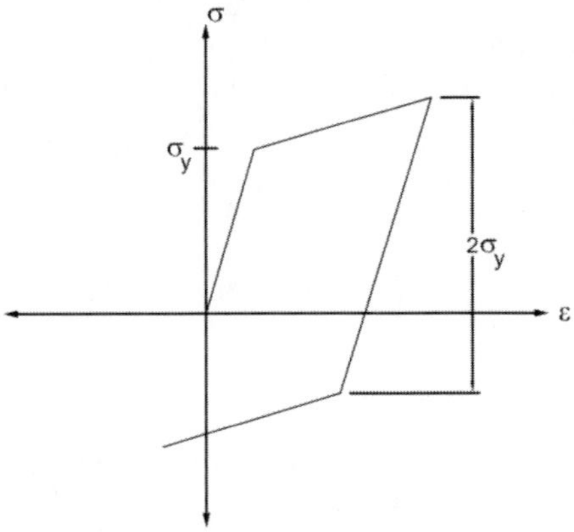

Figure 8. Bauschinger effect.

Creep is a rate dependent material nonlinearity in which the material continues to deform under a constant load. Conversely, if a displacement is imposed, the reaction force (and stresses) will diminish over time (stress relaxation; see Figure 9(a)). The three stages of creep are shown in Figure 9(b).

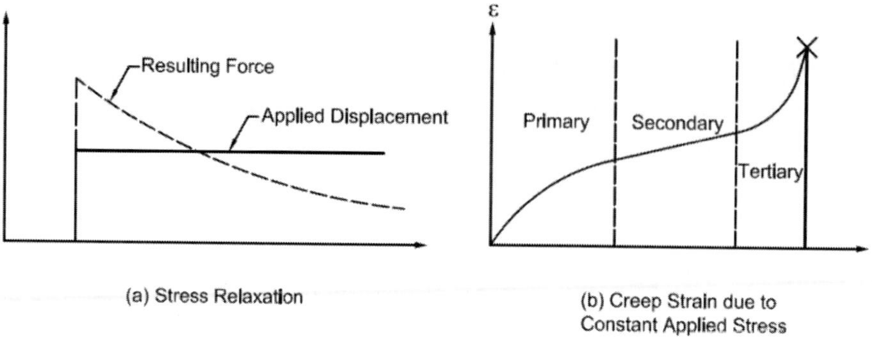

Figure 9. Stress relaxation and creep.

4. Geometry Nonlinearity

Small deflection and small strain analyses assume that displacements are small enough that the resulting stiffness changes are insignificant. Largestrain analyses account for the stiffness changes that result from changes in an element's shape and orientation. All nonlinear effects of initial stresses, large displacements and cable sags are due to geometric nonlinearities.

Deflection Theory of Long-Span Bridges

Long-span bridges exhibit large displacements and cable sags, which contribute to the geometric nonlinearities. In 1901, when the Brooklyn Bridge was designed, Leon Moisseiff realized that the actual behavior of a suspension bridge could be better explained by the deflection theory, a nonlinear analysis, as opposed to the elastic theory. The equation shown below, in terms of h and v, is the fundamental differential equation of the suspension bridge in its classical form.

$$EI.\frac{d^4v}{dx^4}-(H+h)\frac{d^2v}{dx^2}=p+h\frac{d^2y}{dx^2} \tag{4}$$

where v is the deflection of the girder, p is the live loading, h is the cable horizontal component due to p and H is due to the initial tension.

In the fundamental equation of deflection theory, in calculating how the horizontal deck and curved cables worked together to carry loads, the horizontal tension h due to p will be small compared with the initial tension H. Hence, neglecting h on the left side to linearize the equation, the equation can be stated as,

$$EI\frac{d^4v}{dx^4}-H\frac{d^2v}{dx^2}=p+h\frac{d^2y}{dx^2} \tag{5}$$

Therefore, the basic equation becomes a linear differential equation with h directly proportional to p. As a result, v is also proportional to p, and the principle of superposition and the method of influence lines become applicable. Figure 10 shows the deflection–load ratio relations among the elastic, deflection, and linearized deflection theories (Bleich, F. et al., 1950). When the ratio of live load to dead load is small, linearized theory is especially effective for analysis.

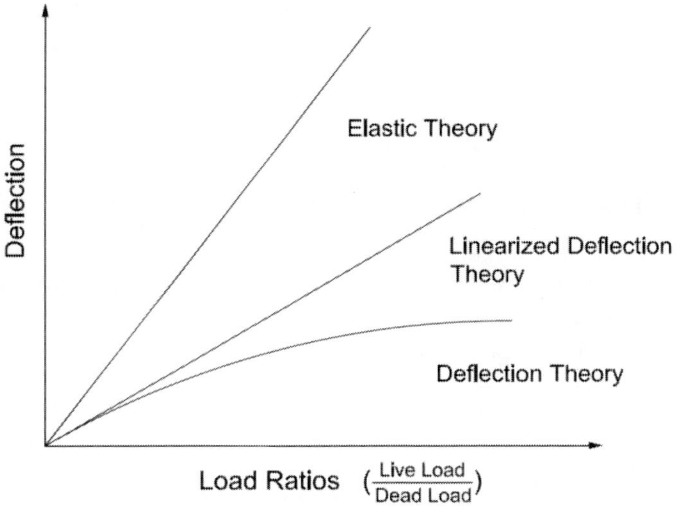

Figure 10. Deflection–load ratios relations among theories.

Modern Matrix Analysis

Nowadays, due to the rapid development of modern day computer and software, matrix analysis can be used for solving nonlinear problems. The actual behavior of the suspension bridge, such as elongation of the hanger ropes, which is disregarded in the deflection theory, can be considered. The nonlinear relationship between force and displacement and its associated geometric nonlinearity, instead of linearization, can be taken care of by iterations. The Newton–Raphson's method, as previously discussed, may be used in these nonlinear matrix displacement analyses for a suspension bridge.

Below, a general geometric nonlinear problem solving principle is illustrated (Zienkiewicz 1977). Despite large displacements of a system, the equilibrium conditions between internal and external 'forces' have to be satisfied as:

$$\psi(a) = \int_V \overline{B}^T \sigma \, dv - f = 0 \tag{6}$$

where ψ is the sum of internal and external generalized forces, a is displacements described at finite number of nodal, f is external forces acting on the structure, σ is stresses, and B defines the relationship between strains and nodal displacements as:

$$\varepsilon = \overline{B}a \tag{7}$$

The bar suffix represents that the relationship is no longer linear since a large displacement is considered, in which strain is now dependent on displacement as

$$\overline{B} = B_0 + B_L(a) \tag{8}$$

If, as assumed, strains are reasonably small, it can still use the general elastic relation as

$$\sigma = D(\varepsilon - \varepsilon_0) + \sigma_0 \tag{9}$$

where D is the elastic matrix representing relationship between the stress and the strain, ε_0 is the initial strain, and σ_0 is the initial stress.

In order to solve a at given f, the Newton-Raphson process is adopted here. Learning the relationship between da and $d\psi$ is required. Taking appropriate variations of equation (6) with respect to da, the follow equation is derived

$$d\psi = (K_0 + K_\sigma + K_L)da = K_T da \tag{10}$$

where K_T is the total tangential stiffness matrix, K_0 represents the usual, small displacements stiffness matrix, i.e.,

$$K_0 = \int_V B_0^T D\, B_0\, dV \tag{11}$$

K_σ is known as initial stress matrix or geometric matrix. It depends on the stress and can be written as

$$\int_V dB_L^T \sigma dV \equiv K_\sigma da \tag{12}$$

K_L is known as the initial displacement matrix or the large displacement matrix and given by

$$K_L = \int_V (B_0^T D\, B_L + B_L^T D\, B_L + B_L^T D\, B_0)dV \tag{13}$$

The procedures of the Newton-Raphson iteration method to solve this nonlinear problem are:

1. obtain the first approximation of the displacement as a^0 based on the elastic linear solution

2. ψ^0, the unbalance force or remained force, is found using equation (6) with appropriatedefinitions of $\overline{B}$ and stress as given by equations (8) and 9
3. K_T^0 is established, and
4. compute the correction of the displacement, Δa^0, by solving the following equation

$$K_T^0 \Delta a^0 + \psi^0 = 0 \tag{14}$$

Repeat (2), (3) and (4) until ψ^n becomes sufficiently small. The loads have to be scheduled step by step when applying above iterations. Otherwise, convergent can hardly be found (Wang and Fu 2012).

It is obvious that the P-Delta effect is reflected in K_σ. When a long cable is affordable to be meshed into small segments, the sag effect will be counted in equation (13). If cables are simply modeled as straight truss elements, Ernst formula (Ernst 1965) is usually adopted to calculate their equivalent stiffness (Tang 1971). However, the drawbacks are that (1) the error will be significant for a long and low stress cable and (2) the coupling of stiffness and stress makes the solving process not unique.

Since the unbalanced forces cannot reach zero exactly, the remainder should be counted when computing ψ^0 in the analysis of successive loads or stage. Otherwise, the error could be significant. The existing state (accumulated displacements and stresses) has to be counted when computing equations (12) and (13) accordingly. Lack of such a feature makes many commercial packages impractical in cable-stayed and suspension bridge analyses.

5. Nonlinearity due to Change of Status

Many common structural features exhibit nonlinear behavior that is status-dependent. For example, a tension-only cable is either slack or taut; a roller support is either in contact or not in contact. Status changes might be directly related to load (as in the case of the cable), or they might be determined by some external cause.

Types of Gap Elements

Gap element is a general term to describe two-node elements formulated in three-dimensional space. Collected from several different software packages capable of handling nonlinear problem, three types of gap elements are available, which are:

1. Bilinear tension-gap-yielding element: This type of element can be used to model cables connected to bridge expansion joints to limit the relative movement during earthquake motions. Figure 11 shows the schematic representation of the link between two nodes and its fundamental behavior.
2. Nonlinear gap-crushing element: This type of element is used to simulate the common type of nonlinear behavior for closing of a gap between different parts of

the structure or the uplifting of the structure at its foundation. It is also used at abutment-soil interfaces and for modeling soil-pile contact during earthquake.

3. 3-D friction-gap element: During the time the surfaces between components of the structures or between structure and foundation are in contact, it is possible for tangential friction forces to develop between the surfaces. If the surfaces are not in contact, the normal and the surface friction forces are zero. In this case, surface slip displacements will take place during the period of time when the allowable friction force is exceeded or when the surfaces are not in "contact".

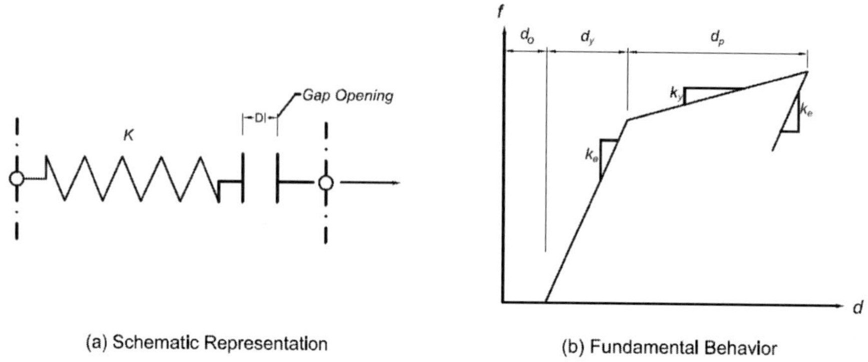

Figure 11. Bilinear tension-gap-yielding element.

Example "Contact" Problems by 3-D Friction-Gap Element

As for the 3-D friction-gap element, situations in which "contact" occurs are common to many different nonlinear applications. "Contact" forms a distinctive and important subset to the category of changing-status nonlinearities.

Contact problems are highly nonlinear and require significant computer resources to solve. It is important to understand the physics of the problem and take the time to set up a model to run as efficiently as possible.

Contact problems present two significant difficulties. First, the regions of contact are generally unknown until running the problem. Depending on the loads, material, boundary conditions, and other factors, surfaces can come into and go out of contact with each other in a largely unpredictable and abrupt manner. Second, most contact problems need to account for friction. There are several friction laws and models to choose from, which are all nonlinear. Frictional response can be chaotic, making solution convergence difficult (ANSYS 1992).

In an example of using the contact element, adjacent precast, prestressed concrete multi-beam bridges transversely post-tensioned studied by Fu et al (2010) is used for demonstration. The ANSYS bridge model of shear keys filled with grout between concrete beams consists of five sections: concrete beams, shear keys, deck, curbs, and transverse post-tension rods (Fu, Pan and Ahmed 2010). Figure 12 illustrates the shape of each section. All the material properties of each section are identical to the properties of the real bridge as well as the completed FEA bridge model.

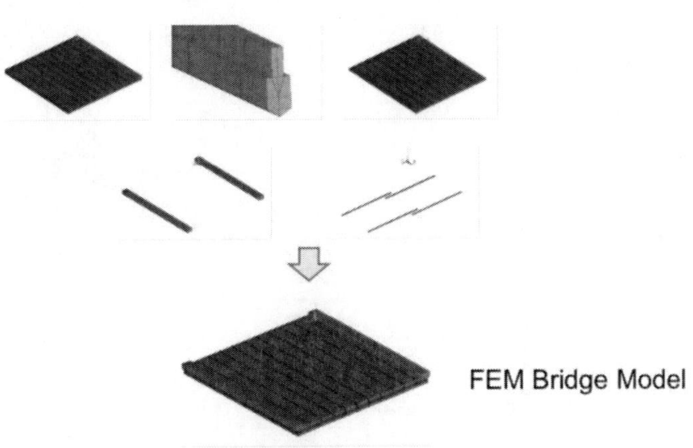

Figure 12. Types of sections in FEM bridge model.

To simulate the crack condition and to avoid overlapping or penetration of elements, the contact element (CONTA174) and target element (TARGE170) were employed to modify the FEA model. These two elements were used between the concrete beams. CONTA174 is brought into play to represent contact and sliding between 3-D target surfaces and deformable surfaces. TARGE170 is discretized by a set of target segment elements and is paired with its associated contact surface via a shared real constant set (ANSYS 1992).

6. Nonlinear Analysis of Long-Span Bridges

Cable Stayed Bridge Example

The first long-span bridge example is a cable stayed bridge (Wang and Fu 2012). Due to its excessive long span, this cable stayed bridge can no longer be assumed as geometric linear. Built in 2008, the longest main span (1088m) in the world at that time, Sutong Bridge crosses over Yanzi River, China. Its feasibility study began in 1999. Figures 13 to 15 show its elevation and main dimensions.

Diaphragms are placed at an interval of 4m, except at anchor positions. Extra counterweights are placed from the first pier to the second auxiliary pier. The analyses focus on only three stages: (1) the service stage, (2) the maximum single-cantilever stage (Figure 16) and (3) the maximum dual-cantilever stage, the following goals are included:

1. Dead loads, live loads and ideal state analyses
2. Static wind load and stability analyses
3. Geometric nonlinear analyses.

The steel girders and towers are modeled as 3D frames, the diaphragms at anchor positions are modeled as rigid bodies and the cables are modeled as 3D trusses.

The FEA model contains 1032 elements and 1035 nodes. For geometric nonlinear analysis, VBDS (Wang and Fu 2005) is employed in the analyses. ANSYS is also used for cross checking. Since the ideal state cannot be defined automatically by cable tuning in

ANSYS (2005), the analysis conducted in ANSYS includes only dead loads analysis in service. Thus, it can only serve as a check.

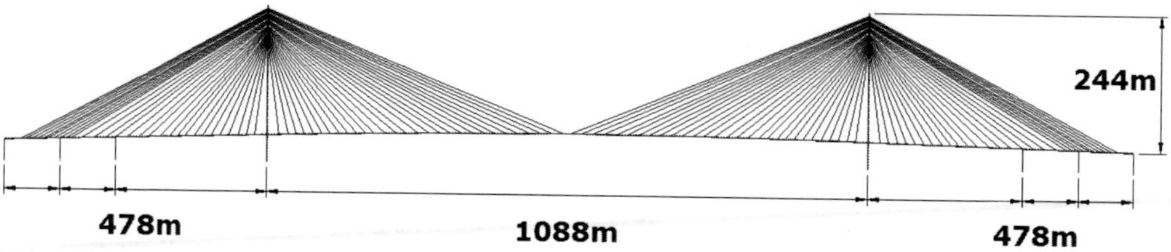

Figure 13. Elevation profile of Sutong Bridge, China.

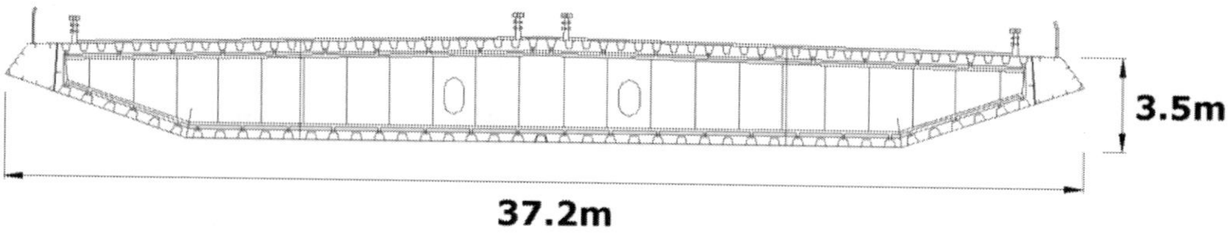

Figure 14. Cross section of Sutong Bridge, China.

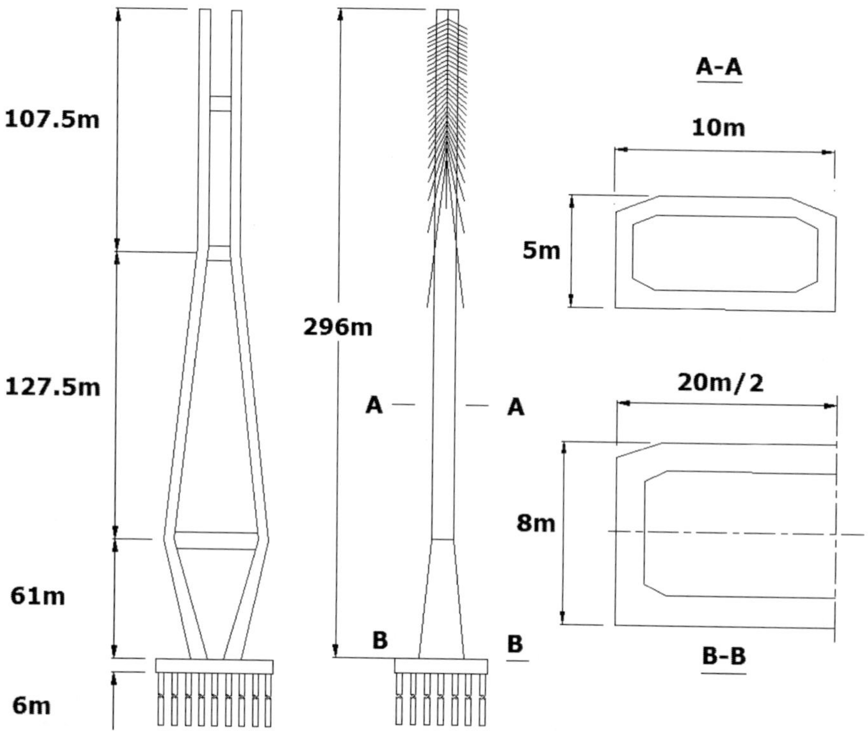

Figure 15. Tower and its cross sections of Sutong Bridge, China.

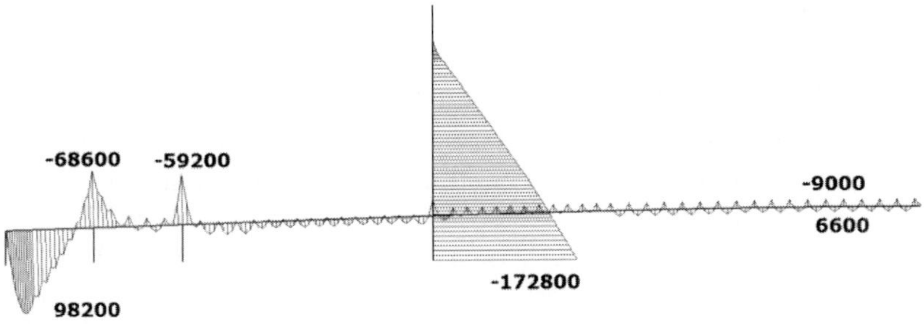

Figure 16. Moment (kNM) distribution of one preferred ideal state.

Table 1 compares the extreme live load displacements with and without geometric nonlinear effects. The nonlinear analysis of live loads shows that the initial stresses accumulated along the flat arch-like girder will increase girder stiffness if large displacement is considered.

Table 1. Live Load Extreme Displacements (mm)

Position	Linear*	Nonlinear**
Girder (in middle of main span)	1081	935
Top of Tower	263	242

* Computed by direct influence line loading, which is obtained by application of unit forces whileinitial stress and sag is considered.

**Obtained by re-analyzing extreme live loads in (1) with consideration of initial stress, cable sag and large displacements.

Suspension Bridge Example

The second example is a suspension bridge, which by nature is a geometrically nonlinear problem (Wang and Fu 2012; Fu, Schwartz, and Mahoney 2007).

The east bound bridge of Chesapeake Bay Bridge, Maryland, USA, 1952, also known as the Bay Bridge, is a suspension bridge with a main span of 487.68m and two 205.74m long suspended side spans. The tower is 107.7m high and the truss stiffened girder passes through it at about its center (Figure 17).

For a geometrically nonlinear problem, VBDS (Wang and Fu 2005) was used to model the bridge in 3D and to perform dead loads and live loads analyses. SAP 2000 (CSI 2009) was used to perform nonlinear blast load analyses. The main cable and hangers are modeled as truss elements; all others are modeled as beam elements. Deck stringers are not modeled, but their weight plus all other superimposed dead loads are included.

All members of each truss floor beam are also modeled as beam elements (Figure 18). The entire model contains 5,264 elements and 2,798 nodes.

The simulation of saddles is critical to the analyses. As frictions do exist between the saddle base and the top of tower, a friction element should be used to simulate the connection between saddle block and tower and nonlinear analysis should be performed, especially for live load analysis.

Two models (Figure 19) could be used to simulate the connection. Due to lack of such an element type in VBDS (Wang and Fu 2005) and the fact that forged steel rollers are used in Bay Bridge for the sliding mechanism, friction forces are simply ignored and the simplified model is used.

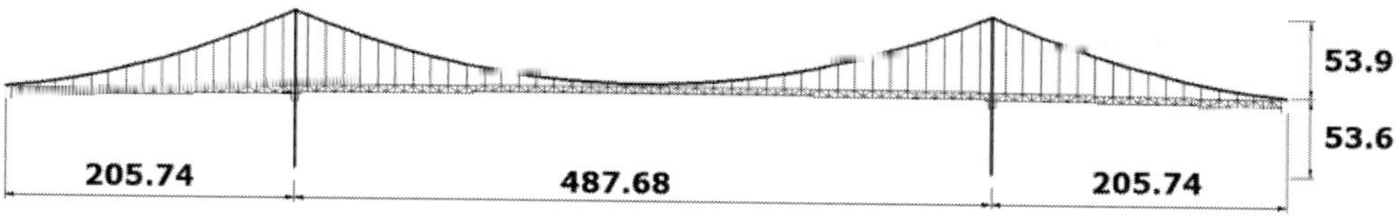

Figure 17. Elevation (m) of Chesapeake Bay Bridge, MD, USA.

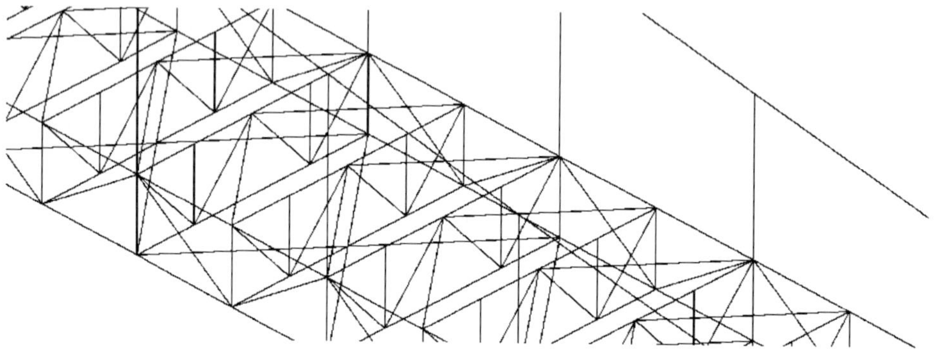

Figure 18. 3D FEA Model of Bay Bridge, USA.

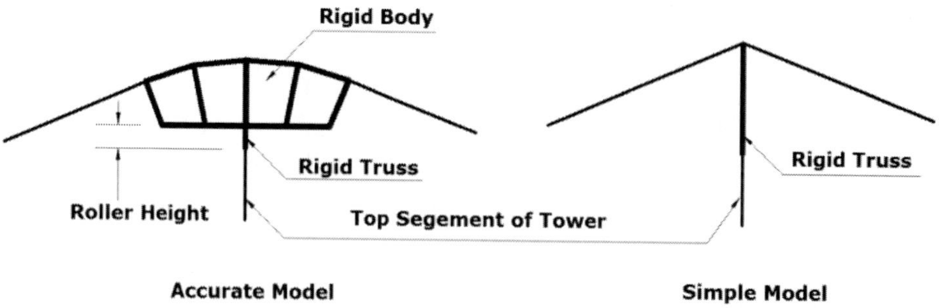

Figure 19. Models for saddle connections.

In addition to dead loads, live loads in main span and far side spans, for extreme bending moment at the bottom of tower leg, and temperature change loads are also analyzed.

As shown in Figures 20 and 21, the extreme bending moment due to live loads is -8,456kNM and its corresponding deflection in the middle of center span is 1,372mm. Table 2 lists additional results of the static analyses.

Figure 20. Extreme live load bending moments (kNM) in towers.

Figure 21. Live load displacements (mm) correspond to extreme bending moments in towers.

Table 2. Static Analyses Results (kN, kNM)

Loads	Cable Reaction	Tower Leg Reaction	
		Axial	Moment
Dead loads	29,304	-31,169	0
Live loads*	7,192	-5,026	-8,456
26 °C temp. drop	555	-147	1,901
11 °C temp. raise	-369	98	-1,268

*3 lanes with each of 0.87kN/m + 80kN concentrated load, lanes discount of 0.9 is used.

Due to nonlinearity, the dead loads analysis is performed first. Several rounds of iterations are needed in order to consider the initial stresses in cables due to dead loads. These initial stresses will be automatically considered in succeeding analyses of live loads and temperature loads.

7. NONLINEAR SEISMIC ANALYSIS OF BRIDGE STRUCTURES

Linear and Nonlinear Seismic Analyses

Reducing earthquake losses in transportation systems begins before earthquakes occur by identifying the bridges that are likely to fail and cause disruption to the system and when to rehabilitating them. Four distinct analytical procedures, as shown in Figure 22, can be used in systematic rehabilitation of structures (FEMA-273, 1997): Linear Static, Linear Dynamic, Nonlinear Static (Pushover), and Nonlinear Dynamic Procedures (NDP). Linearly elastic procedures (linear static and linear dynamic) are the most common procedures in seismic analysis and design of structures due to their simplicity. On the other hand, adjustments to overall deformations and material acceptance criteria can be incorporated to consider the inelastic response and this can be performed by reducing the seismic loading based on the ductility of structures.

A large number of bridges were designed and constructed at a time when bridge codes had no seismic design provisions, or when these provisions were insufficient according to current standards. Many of these bridges may suffer severe damage when struck by earthquakes, as evidenced by recent moderate earthquakes.

Linear elastic procedures are sufficient as long as the structure behaves within elastic limits. If the structure responds beyond the elastic limits, linear analyses may indicate the location of first yielding, but cannot predict failure mechanisms and account for redistribution of forces during progressive yielding. This fact makes the elastic procedures insufficient to perform assessments and retrofitting evaluations for those bridges in particular and structures in general. Nonlinear (static and dynamic) procedures are the solutions that can overcome this problem and show the performance level of the structures under any loading level.

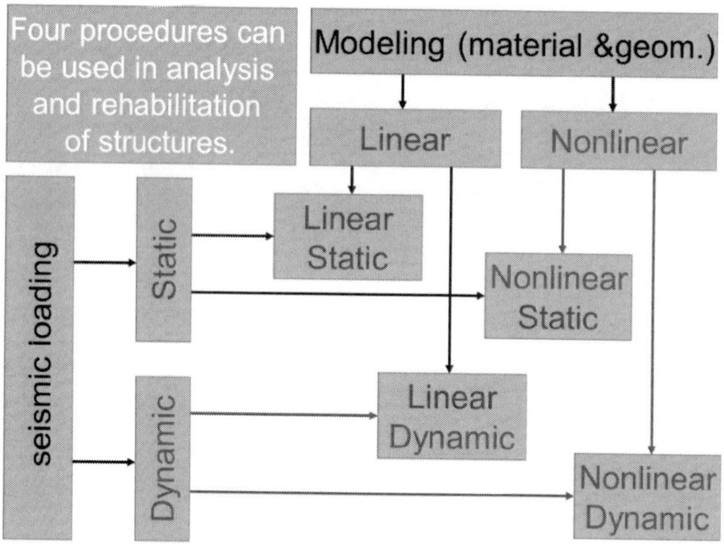

Figure 22. Four distinct analytical procedures for seismic analysis.

Nonlinear procedures can also help demonstrate how structures really work by identifying modes of failure and the potential for progressive collapse. Nonlinear procedures will help engineers to understand how a structure will behave when it is subjected to major earthquakes, assuming that the structure will respond beyond the elastic limits, and this will resolve some of the uncertainties associated with codes and elastic procedures.

The performance approach, which was proposed in AASHTO Guide Specification (2009), will be considered, as shown in Table 3.

Table 3. Performance Approach

Probability of Exceedence For Design Earthquake Ground Motions		Performance Level	
		Life Safety	Operational
Rare Earthquake (MCE) 3% in 75 years	Service	Significant disruption	Immediate
	Damage	Significant	Minimal
Frequent of Expected Earthquake 50% in 75 years	Service	Immediate	Immediate
	Damage	Minimal	Minimal to none

Performance-based engineering, with their performance levels shown in Figure 23, is set to select design structural criteria such that at specified level ground motion, the structure will not be damaged beyond certain limiting states.

In this article, conventional dynamic analysis (nonlinear dynamic in Figure 22) and modal pushover analysis procedures (nonlinear static in Figure 22) to determine seismic demands for inelastic structures are presented.

Nonlinear Time-History Analysis

Time-history analysis is a step-by-step analysis of the dynamical response of a structure to a specified loading that may vary with time through a process of numerical integration of the equations of motion. It involves the development of a complete mathematical model of the bridge wherein an effort is made to model nonlinear forms of behavior in a highly localized (rather than global) manner. The mathematical model is formulated in such a way that the stiffness and even connectivity of the elements can be directly modified based on the deformation state of the structure. This permits the effects of element yielding, buckling, and other nonlinear behavior on structural response to be directly accounted for in the analysis. The model is then subjected to time-histories of earthquake ground acceleration that may be in either historical records or design spectrum compatible records. In either case, an attempt is made to capture the full time-history of the nonlinear structural response.

Nonlinear dynamic analysis has some problems for routine application to typical highway bridges. For example, there are difficulties in developing reasonable models for all the nonlinear components of the bridge, the results often exhibit strong sensitivity to the details of the model, and the extensive output can be difficult to interpret while masking important aspects of the seismic response. Another limit on the accuracy of this approach is the fact that minor deviations in ground motion or even in element hysteretic behavior can result in significant differences in predicted response of structures (Aviram, Mackie, and Stojadinovic 2008). For these reasons, when nonlinear response history analysis is used in the design process, it is necessary to run a suite of ground motion records.

The use of multiple records in the analyses allows observation of the difference in response resulting from differences in record characteristics. As a minimum, suites of ground motions include at least three different records are recommended, (FEMA-450).

Different from linear time history analysis, the differential equations of motion for nonlinear time history analysis cannot be considered as smooth functions. It is due to the nonlinear hysteresis of most bridge structural materials, friction forces developed between contacting surfaces, and buckling of elements.

Therefore, only step-by-step methods are recommended for the solution of the nonlinear time history of bridge structures. The step-by-step solution methods attempt to satisfy dynamic equilibrium at discrete time steps and may require iteration, especially when nonlinear behavior is developed in the structure and the stiffness of the complete structural system must be recalculated due to degradation of strength and redistribution of forces (Aviram, Mackie, and Stojadinovic 2008).

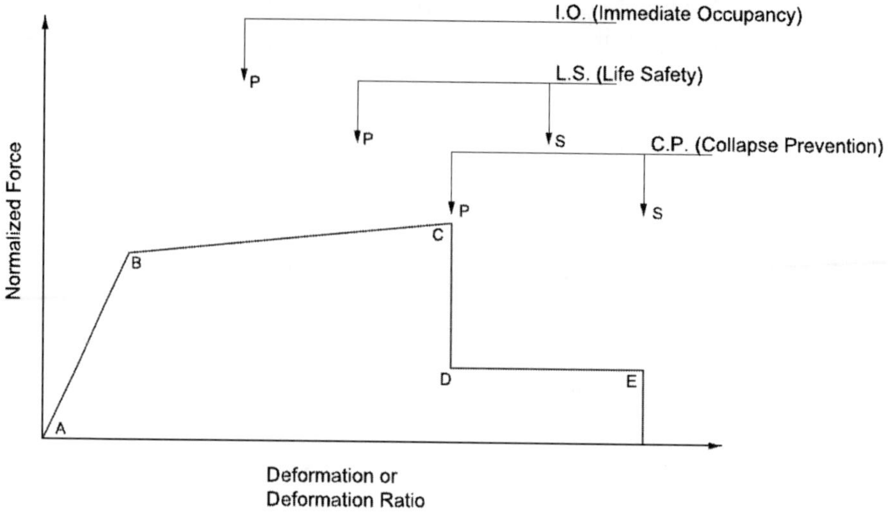

Figure 23. Performance level of structures.

Unlike linear time history, the nonlinear case can take a significant amount of time to solve structural systems with just a few hundred degrees of freedom.

Engineers must be very careful in the interpretation of the results and check the results using the applicable acceptance criteria. An example of a bridge case analyzed by linear and nonlinear time history analysis methods plotted on the same graph is shown in Figure 24.

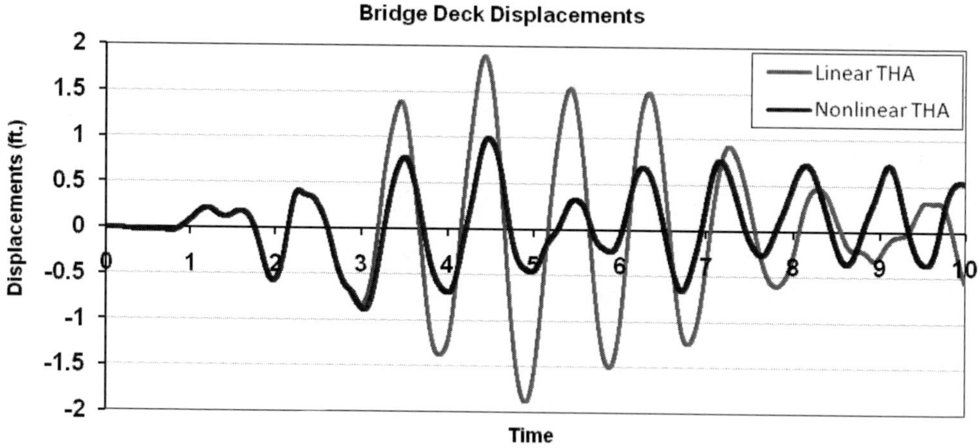

Figure 24. Linear vs. Nonlinear time history analysis for a 9-Span bridge model.

Modal Pushover Analysis (MPA) Procedure for Bridges' Application

Nonlinear static analysis (Pushover) is widely used as a design tool for the assessment of the structural behavior in the inelastic range and identifying the locations of structural weaknesses and failure mechanisms.Yet it is limited to structures where responses are controlled by the fundamental mode. On the other hand, Modal Pushover Analysis (MPA) method has been presented by Chopra and Goel (2002) for complex building structures which accounts for higher mode effects on the behavior of structures. Due to the nature of bridges, which extend horizontally, rather than buildings which extendvertically, some considerations and modifications should be taken into consideration to render the MPA applicable for bridges. Key elements of applying the MPA procedure for the case of bridges are:

- Definition of the control node: The control node is used to monitor displacement of the structure. Its displacement versus the base-shear forms the capacity (pushover) curve of the structure.
- Development of the pushover curve and transformation it into a capacity curve.
- Use of the capacity spectrum for defining the earthquake demand for each mode.
- The number of modes that should be considered in the case of bridges.

Step-by-step extended MPA procedure for bridges was proposed (Ahmed 2012; Ahmed and Fu 2012) and is listed here:

1. Compute the natural periods, T_n and modes ϕ_n, for linearly elastic vibration of the structure.
2. Carry out separate pushover analyses for force distribution, $S_n^* = m\phi_n$, where m is the mass matrix of the structure, for each significant mode of the bridge, and construct the base shear versus displacement of the monitoring point (V_{bn}–u_{rn}) pushover curve for each mode. Gravity loads are applied before each MPA, and P-Δ effects are included, if significant (e.g. bridges with tall piers). It is noted that the value of the lateral deck displacement due to gravity loads, u_{rg}, is negligible for a bridge with nearly symmetrically distributed gravity loading.
3. Idealize the pushover curve as a bilinear curve so that a yield point and ductility factor can be defined and then used to appropriately reduce the elastic response spectra representing the seismic action considered for assessment.
4. This idealization can be done in a number of ways, some more involved than others; it is suggested to do this once as recommended by Paraskeva et al. (2006) (as opposed, for instance, to the ATC-40 1996 procedure) using the full pushover curve (i.e. analysis up to 'failure' of the structure, defined by a drop in peak strength of about 20%) and the equal energy absorption rule (equal areas under the actual and the bilinear curve). Remaining steps of the MPA procedure can be applied even if a different method for producing a bilinear curve is used.
5. Convert the idealized pushover curve (V_{bn}–u_{cn}) of the multi-degree-of freedom, MDOF, system (calculated in Step 3) to a capacity diagram, as shown in Figure 25 Right. The base shear forces and the corresponding displacements in each pushover curve are converted to spectral accelerations (S_a) and spectral displacements (S_d), respectively, of an equivalent single degree-of-freedom (SDOF) system, using the relationships (Chopra and Goel 2002, ATC-40 1996):

$$S_a = \frac{V_{bn}}{M_n^*} \tag{15}$$

$$S_d = \frac{u_{cn}}{\Gamma_n \phi_{cn}} \tag{16}$$

wherein ϕ_{cn} is the value of the mode shape ϕ_n at the monitoring point, $M_n^* = L_n * \Gamma_n$ is the effective modal mass, $L_n = \phi_n^t m.1$, $\Gamma_n = L_n / M_n$, and $M_n = \phi_n^t m \phi_n$ is the generalized mass, for the nth natural mode. For inelastic behavior, inelastic spectrum is used for estimating the displacement demand at the monitoring point. The formula proposed by Fajfar (1999) was used.

6. Convert of the displacement demand of the *n*th mode inelastic SDOF system to the peak displacement of the monitoring point, u_{cn} of the bridge, using equation (16).
7. If the structure remains elastic or close to the yield point, no correction is needed for the demand displacement calculated in the previous step. For cases in whichsignificant inelasticity develops in the structure, a correction is made to the displacement of the monitoring point of the bridge to estimate the modified control point displacement u'_{cn}. The response displacements of the structure are evaluated by extracting from the database of the individual pushover analyses the values of the desired responses at which the displacement at the control point is equal to u_{cn} (see Eq. 16). These displacements are then applied to derive a new vector ϕ_n', which is the deformed shape (affected by inelastic effects) of the bridge subjected to the given modal load pattern. The target displacement at the monitoring point for each pushover analysis is calculated again with the use of ϕ_n', according to:

$$u'_{cn} = \Gamma'_n \cdot \phi'_{cn} \cdot S_{dn} \tag{17}$$

wherein S_{dn} is the displacement of the SDOF system and Γ_n' is Γ_n recalculated using ϕ_n'.

8. The response quantities of interest (displacements, plastic hinge rotations, forces in the piers) are evaluated by extracting from the database of the individual pushover analyses the values of the desired responses r_n, due to the combined effects of gravity and lateral loads for the analysis step at which the displacement at the control point is equal to u'_{cn} (see Eq. 17).
9. Steps 3 to 7are repeated for as many modes as required for sufficient accuracy.
10. The total value for any desired response quantity (and each level of earthquake intensity considered) can be determined by combining the peak 'modal' responses, r_{no}, using an appropriate modal combination rule (e.g. the SRSS combination rule). This simple procedure was used for displacements, total base shear and plastic hinge rotations in the present study, which were the main quantities used for assessing the bridges analyzed (whose response to service gravity loading was, of course, elastic)

Due to the Simplicity of applying the MPA procedure without the need to obtain different ground motion records, the main obstacle to dynamic analysis, MPA might be viewed as an upper-bound level of sophistication for nonlinear static analysis and is more favorable as a design tool in the current practice.

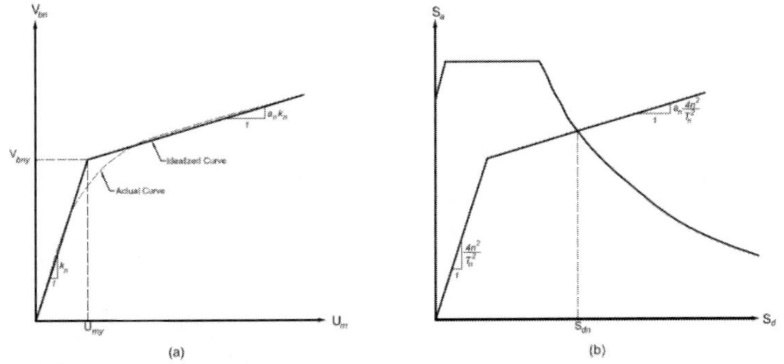

Figure 25. Idealized pushover curve of the nth mode of the MDOF system, and corresponding capacity curve for the nth mode of the equivalent inelastic SDOF system.

Case Study

1. Analyzed Bridge Model

This bridge is one of the FHWA examples series (FHWA, 1996-a; Ahmed 2010). It consists of three spans as shown in Figure 26. The total length is 320 feet with span lengths of 30, 36, and 30 meters (100, 120, and 100 feet), respectively. All substructure elements are oriented at a 30-degree skew from a line perpendicular to a straight bridge centerline alignment. The superstructure is a cast-in-place concrete box girder with two interior webs, as shown in Figure 27. The intermediate bents have a crossbeam integral with the box girder and two round columns that are pinned at the top of spread footing foundations. SAP2000 program (CSI 2009) was used to perform the analysis. Figure 28 shows the finite element model of the bridge. Table 4 lists the section properties used for the finite element model of the bridge.

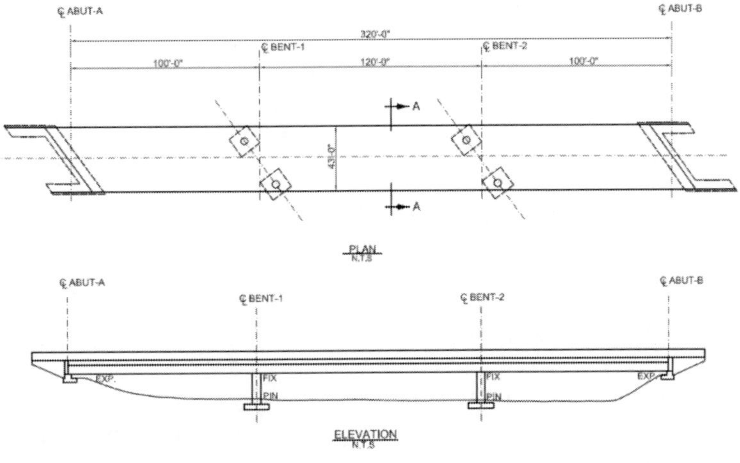

Figure 26. Bridge model – plan and elevation.

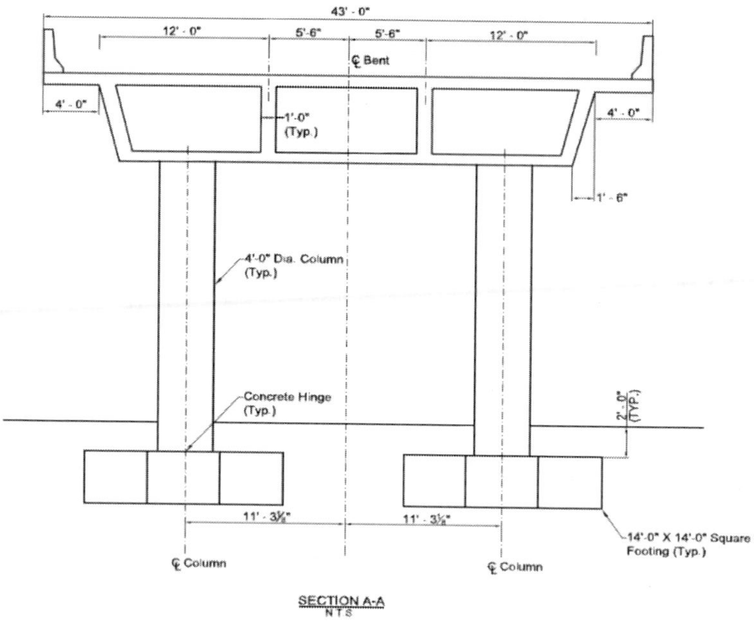

Figure 27. Bridge model – typical cross section.

2. Foundation Stiffness

The intermediate bent foundations were modeled with equivalent spring stiffness for the spread footing. For this bridge, all of the intermediate bent footings use the same foundation springs. The stiffness values are developed for the local bent supports and transformedto global support when input to SAP2000 program (CSI 2009), so as to have compatible results for the MPA analysis and the nonlinear time-history analysis.

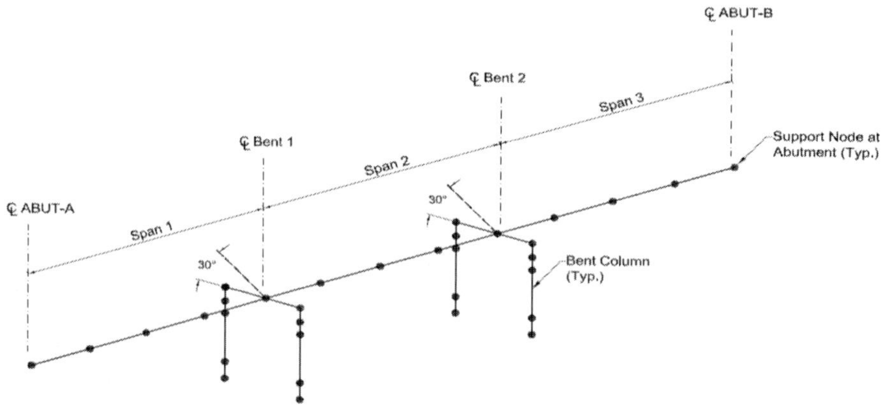

Figure 28. Bridge model – finite element model.

Values of stiffness for foundation springs provided by FHWA (1996-a) are used in this study.The abutments have been modeled with a combination of full restraints (vertical translation and superstructure torsional rotation) and equivalent spring stiffness (transverse translation); other degrees of freedom are all released.Tables 5 to 7 list the dynamic characteristics required within the context of the MPA analysis (Ahmed 2012; Ahmed and Fu 2012).

Table 4. Section Properties for the Bridge Model (Ahmed 2010)

Element Properties	CIP Box Superstructure	Bent Cap Beam	Bent Column
Area (ft^2)	72.74	27.00	12.57
I_x – Torsion (ft^4)	1177	100000 (1)	25.13
I_y – (ft^4)	401	100000 (2)	9.00
I_z – (ft^4)	9697	100000 (3)	9.00

Notes: This value has been increased for force distribution to bent columns. Actual value is $I_x = 139\ ft^4$
This value has been increased for force distribution to bent columns. Actual value is $I_y = 90\ ft^4$
This value has been increased for force distribution to bent columns. Actual value is $I_z = 63\ ft^4$

Table 5. Modal Periods and Frequencies

Output Case	Step Type	Step Num	Period Sec	Frequency Cyc/sec	CircFreq rad/sec	Eigenvalue rad2/sec2
Modal	Mode	1.000000	0.966007	1.0352E+00	6.5043E+00	4.2306E+01
Modal	Mode	2.000000	0.526100	1.9008E+00	1.1943E+01	1.4263E+02
Modal	Mode	3.000000	0.210878	4.7421E+00	2.9795E+01	8.8777E+02
Modal	Mode	4.000000	0.125163	7.9896E+00	5.0200E+01	2.5200E+03
Modal	Mode	5.000000	0.081535	1.2265E+01	7.7061E+01	5.9385E+03
Modal	Mode	6.000000	0.068764	1.4543E+01	9.1373E+01	8.3491E+03
Modal	Mode	7.000000	0.048488	2.0624E+01	1.2958E+02	1.6792E+04
Modal	Mode	8.000000	0.034272	2.9178E+01	1.8333E+02	3.3610E+04
Modal	Mode	9.000000	0.030670	3.2605E+01	2.0486E+02	4.1969E+04
Modal	Mode	10.000000	0.024273	4.1198E+01	2.5885E+02	6.7004E+04
Modal	Mode	11.000000	0.022045	4.5361E+01	2.8501E+02	8.1233E+04
Modal	Mode	12.000000	0.018333	5.4547E+01	3.4273E+02	1.1747E+05

Table 6. Modal Participation Factors

Output Case	Step Type	StepNum	Period Sec	UX Kip-s2	UY Kip-s2	UZ Kip-s2
Modal	Mode	1.000000	0.966007	11.908197	0.254998	0.000000
Modal	Mode	2.000000	0.526100	0.273393	-11.131288	0.000000
Modal	Mode	3.000000	0.210878	-1.907E-12	-9.051E-14	0.000000
Modal	Mode	4.000000	0.125163	-0.001474	-4.179216	0.000000
Modal	Mode	5.000000	0.081535	1.396E-11	2.655E-13	0.000000
Modal	Mode	6.000000	0.068764	-3.028E-11	-8.692E-13	0.000000
Modal	Mode	7.000000	0.048488	0.000825	0.654170	0.000000
Modal	Mode	8.000000	0.034272	0.008127	-0.000365	0.000000
Modal	Mode	9.000000	0.030670	7.547E-11	1.627E-12	0.000000
Modal	Mode	10.000000	0.024273	5.445E-10	1.165E-11	0.000000
Modal	Mode	11.000000	0.022045	-0.000359	0.122636	0.000000
Modal	Mode	12.000000	0.018333	0.004605	-0.000223	0.000000

Figure 29 presents an example of the pushover curve obtained using the second mode shape as the lateral load pattern (mode shape multiplied by the mass).

Table 7. Modal Participating Mass Ratios

Output Case	Step Type	StepNum	Period	UX	UY	UZ	SumUX	SumUY
			Sec					
Modal	Mode	1.000000	0.966007	0.99947	0.00046	0.00000	0.99947	0.00046
Modal	Mode	2.000000	0.526100	0.00053	0.87331	0.00000	1.00000	0.87377
Modal	Mode	3.000000	0.210878	0.00000	0.00000	0.00000	1.00000	0.87377
Modal	Mode	4.000000	0.125163	1.531E-08	0.12310	0.00000	1.00000	0.99687
Modal	Mode	5.000000	0.081535	0.00000	0.00000	0.00000	1.00000	0.99687
Modal	Mode	6.000000	0.068764	0.00000	0.00000	0.00000	1.00000	0.99687
Modal	Mode	7.000000	0.048488	4.794E-09	0.00302	0.00000	1.00000	0.99989
Modal	Mode	8.000000	0.034272	4.655E-07	9.366E-10	0.00000	1.00000	0.99989
Modal	Mode	9.000000	0.030670	0.00000	0.00000	0.00000	1.00000	0.99989
Modal	Mode	10.000000	0.024273	0.00000	0.00000	0.00000	1.00000	0.99989
Modal	Mode	11.000000	0.022045	9.086E-10	0.00011	0.00000	1.00000	1.00000
Modal	Mode	12.000000	0.018333	1.494E-07	3.512E-10	0.00000	1.00000	1.00000

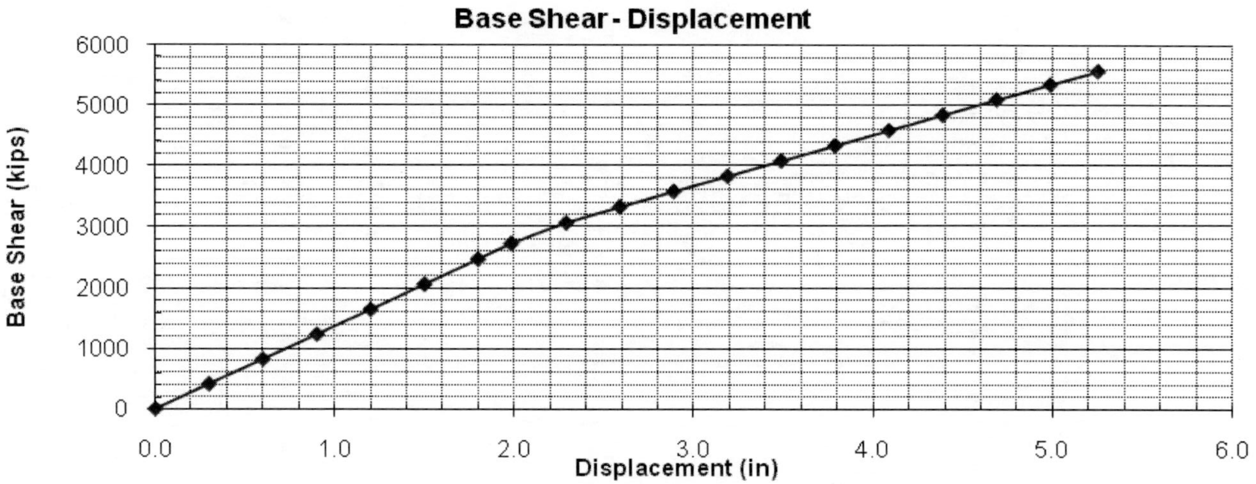

Figure 29. Pushover curve for the second mode shape load pattern.

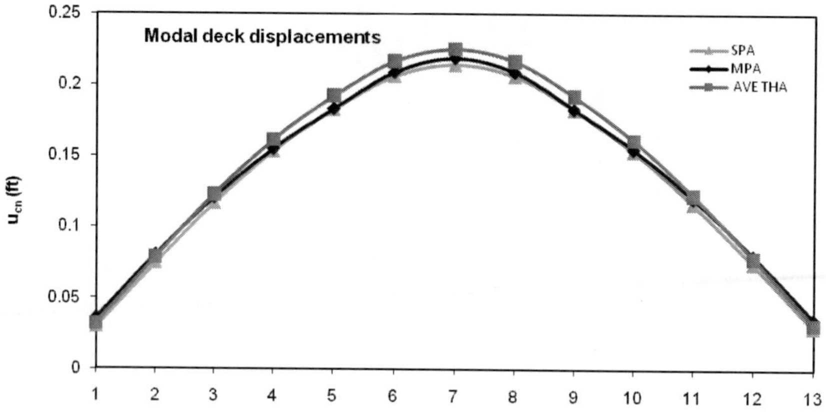

Figure 30. Deck displacements for bridge model calculated from SPA, MPA and THA, For PGA = 0.30g.

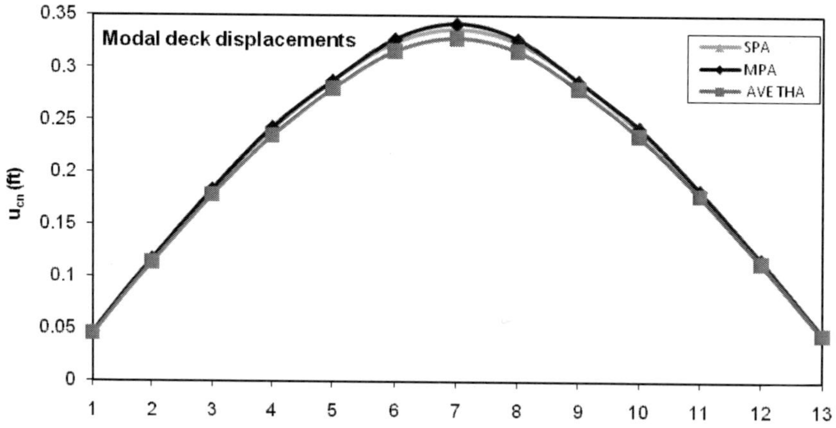

Figure 31. Deck displacements for bridge model calculated from SPA, MPA and THA, For PGA = 0.45g.

Nonlinear time-history analyses (NL-THA) were also performed using three different ground acceleration histories and the average was taken.The average response was compared with those from the modal pushover analysis and the standard pushover analysis using different values of PGA (0.30g and 0.45g) for the design response spectra. Figures 30 and 31 show deck displacements obtained using different analysis methods.

It is observed that the MPA predicts well the transverse displacements of the bridge with only 3% difference from those results obtained from the time-history analysis. As the level of excitation increases, the displacement profile derived by the MPA tends to match that obtained from the NL-THA. As for the base shear forces, MPA predicted a total base shear of 3059.06 kips for the first level of earthquake excitation (PGA=0.30g) compared to 2983.02 kips obtained from the NL-THA case with a difference of only 2.5%. On the other hand, for PGA=0.45g, a base shear value of 4124.8 kips was predicted compared to a value of 3877.23 kips from NL-THA with a difference of 6.4%.

CONCLUSION

Many computer programs are available for solving bridge problems. Their mathematical models include loads, geometry and material, as well as the methodology for bridge analysis to check against their limit states where limit states are defined as conditions of a structure beyond which it no longer fulfills the relevant design criteria. Based on the latest AASHTO LRFD Specifications (2012), four types of limit states are considered: (1) strength limit states, (2) service limit states, (3) fatigue limit states, and (4) extreme limit states. Service and fatigue limit states are considered under elastic conditions and analyzed as fully elastic and linear for all cases. Strength limit state, even with the load factors employed, can be analyzed as an elastic system, except in the case where inelastic redistribution of negative bending moment for a continuous girder system. The extreme limit states, on the other hand, may require collapse investigation based entirely on inelastic and nonlinear modeling (AASHTO 2012). As stated in the AASHTO (2012), for long-span bridges which are very flexible, such as suspension and cable-stayed bridges, nonlinear elastic methods, such as large deflection theory, should be used. Also considered as nonlinear problems are changes of state, such as support lift-off, and contact problems.

Within the four limit states, strength and service limit states are considered in the design process. In the checking process, fatigue limit states are analyzed elastically while extreme limit states are analyzed inelastically for most cases. Nowadays, with the rapid development of modern day software, many sophisticated structural programs can be used for solving nonlinear problems under extreme limit states to complete the bridge design and checking process. Furthermore, nonlinear seismic analysis procedures help designers to understand how bridge structure behaves when it is subjected to major earthquakes. With these tools, nonlinear problem solving becomes more or less a routine process for structural engineers. Through this article, engineers may become familiar with those complexities and assumptions to analyze bridges in the nonlinear range to make them safe and economical.

REFERENCES

American Association of State Highway and Transportation Officials (AASHTO).2012. *AASHTO LRFD Bridge Design Specifications*. 6th ed. Washington, DC: AASHTO.

AASHTO 2009.*AASHTO Guide Specification for LRFD Seismic Bridge Design.* American Association of State Highway and Transportation Officials.

Ahmed, M. S. 2010. Seismic assessment of curved bridges using modal pushover analysis. *PhD Dissertation.*College Park, MD: Department of Civil and Environmental Engineering, University of Maryland.

Ahmed, M.S. and Fu, C.C., 2012. "Seismic Assessment of Long Curved Bridges using Modal Pushover Analysis: A Case Study," the *Proceedings of the 6th International Conference on Bridge Maintenance, Safety and Management (IABMAS2012)*, July 8-12, 2012, Como, Italy.

ANSYS users' manual for revision 10.0 (2005); theoretical manual, Part I. Elementary Library, Houston: Swanson Analysis Systems Inc.; 1992.

Aviram, A.; Mackie, K. R.; Stojadinovic, B.2008, *Guidelines for nonlinear analysis of bridge structures in California*, PEER-2008/03, Pacific Earthquake Engineering Research Center (PEER), University of California, Berkeley, CA.

Bleich, F. et al., 1950. *The Mathematical Theory of Vibrations in Suspension Bridges*, Bureau of Public Roads, Washington, D.C.

Chopra, A. K., and Goel, R. K. 2002. A Modal pushover analysis procedure for estimating seismic demands for buildings. *Earthquake Engineering and Structural Dynamics, 31*(3), 561-582.

Chopra, A. K., and Goel, R. K. 2004. A modal pushover analysis procedure to estimate seismic demands for unsymmetrical-plan buildings. *Earthquake Engineering and Structural Dynamics, 33*(8), 903-927.

CSI. 2009. SAP 2000 Advanced version 14.0. Berkeley, CA: Computers and Structures, Inc.

Ernst, J. H. 1965. Der E-Modul von Seilenunterberucksichtigung des Durchhanges. *Der Bauingenieur* 40(2):52-55.

Fajfar, P., and Fischinger, M. 1989. N2 - A method for non-linear seismic analysis of regular buildings. *The 9th World Conference on Earthquake Engineering*, *V*, pp. 111-116. Tokyo-Kyoto, Japan.

Federal Emergency Management Agency (FEMA) (1997), "NEHRP Guidelines for the Seismic Rehabilitation of Buildings, "*FEMA-273*, Washington, D.C.

Federal Emergency Management Agency (FEMA) (1997). "NEHRP Recommended Provisions for Seismic Regulations for New Buildings and Other Structures."*FEMA-302*, Washington, D.C.

Fedral Emergency Management Agency (FEMA)-. (2000). "Prestandard and Commentary for the Seismic Rehabilitation of Buildings, "*FEMA-356*, Washington, D.C.:.

Federal Emergency Management Agency (FEMA) (2003). "NEHRP Recommended Provisions for Seismic Regulations for New Buildings and Other Structures. "*FEMA-450*, Washington, D.C.

FHWA 1996-b. 1996. *Seismic Design of Bridges: Design Example No. 5 Nine span Viaduct Steel Girder Bridge.* Springfield, VA: Fedral Highway Administration; Publication No. FHWA-SA-97-009.

Fu, C.C., Pan, Z.F., and Ahmed, M.S., "Transverse Post-tensioning Design of Adjacent Precast Solid Multi-beam Bridges," *ASCE Journal of Performance for Constructed Facilities*, June 2010.

Fu, C.C., Schwartz, C.W., and Mahoney, E., "Phase II Report – Highway Cost Allocation Study for the William Preston Lane, Jr. Memorial (Bay) Bridge," *Technical Report to Maryland Transportation Authority* (MDTA), 2007.

Fu, C.C. and Alayed, H., "Seismic Analysis of Bridges Using Displacement-Based Approach," (03-3175) the *Proceedings of Transportation Research Board*, January 12-16, 2003, Washington, D.C.

Hognestad, E. (1951). "A study on combined bending and axial load in reinforced concrete members." Univ. of Illinois Engineering Experiment Station, Univ. of Illinois at Urbana-Champaign, IL, 43-46.

Kent, D.C., and Park, R. (1971)."Flexural members with confined concrete". *Journal of the Structural Division, Proc. of the American Society of Civil Engineers,* 97(ST7), 1969-1990.

Mander, J.B., Priestley, M.J., and Park, R. (1988a) "Observed stress-strain behavior of confined concrete". *J. Struct. Eng.,* 114(8), 1827-1849.

Mander, J.B., Priestley, M.J.N., and Park, R. (1988b) "Theoretical stress-strain model of confined concrete". *J. Struct. Eng.,*114(8), 1804-1826.

Tang, M.C. 1971. Analysis of cable-stayed girder bridges. *Journal of the Structural Division, ASCE* 97(ST5):1482-1495.

Tang, M.C. 1976. Bulking of cable-stayed bridges. *Journal of the Structural Division, ASCE* 102(ST7):1675-1684.

Wang, S., and Fu, C.C., "Structural Design and Analysis of Long Span Bridges." the *Proceedings of the 6th International Conference on Bridge Maintenance, Safety and Management (IABMAS2012),* July 8-12, 2012, Como, Italy.

Wang, S. Q. and Fu, C.C., 2005. *VBDS*, Visual Bridge Design System Version 1.1.

Zienkiewicz, Olgierd Cecil 1977. *The Finite Element Method (Third Edition).* McGraw-Hill Book Company (UK) Limited.

In: Focus on Nonlinear Analysis Research
Editors: Gualtiero Padovani and Manuel Occhino
ISBN: 978-1-62417-350-9
© 2013 Nova Science Publishers, Inc.

Chapter 4

NONLINEAR ANALYSIS OF SPONTANEOUS MEG ACTIVITY IN ALZHEIMER'S DISEASE

Carlos Gómez*, Jesús Poza, María García, Alberto Fernández and Roberto Hornero
Biomedical Engineering Group, E.T.S. Ingenieros de Telecomunicación,
University of Valladolid, Spain

ABSTRACT

Alzheimer's disease (AD) is one of the most common disorders among elderly population and it is considered the main cause of dementia in western countries. This progressive, degenerative brain disorder is characterized by neural loss and the appearance of neurofibrillary tangles and senile plaques. Although a definite diagnosis is only possible by necropsy, a differential diagnosis with other types of dementia and with major depression should be attempted. The differential diagnosis includes medical history studies, physical and neurological evaluation, mental status tests, and neuroimaging techniques. Nowadays, magnetoencephalography (MEG) recordings are not used in AD clinical diagnosis, despite its potential as aid diagnostic tool.

The aim of this chapter is to show recent analyses of the spontaneous MEG activity in AD. For this purpose, five minutes of recording were acquired with a 148-channel whole-head magnetometer in 36 patients with probable AD and 26 control subjects. MEG data were analyzed by means of three nonlinear measures: Lempel-Ziv complexity (*LZC*), sample entropy (*SampEn*) and detrended fluctuation analysis (*DFA*). *LZC* measures the number of different substrings and the rate of their recurrence along the original time series. *SampEn* is an embbeding entropy that quantifies the signal regularity. Finally, *DFA* is a method designed to quantify correlations in noisy and non-stationary time series.

LZC and *SampEn* showed that MEG recordings are less complex and more regular in AD patients than in control subjects. *DFA* revealed changes in the fluctuations of MEG signals. Significant differences between AD patients and elderly controls were found

* Corresponding author: Biomedical Engineering Group, E.T.S. Ingenieros de Telecomunicación, University of Valladolid, Paseo Belén, 15, 47011 Valladolid (Spain), Telephone: +34 983423981, Fax: +34 983423667, E-mail: cargom@tel.uva.es, Web-site: www.gib.tel.uva.es.

using the three nonlinear methods (p-values < 0.05, Welch t-test). We used receiver operating characteristic (ROC) curves with a leave-one-out cross-validation procedure to assess the ability of the methods to classify AD patients and control subjects. The highest area under the ROC curve was achieved with *DFA* (0.7959) whereas the highest accuracy was obtained with *LZC* algorithm (72.58%).

We conclude that nonlinear analyses from spontaneous MEG activity could be useful to increase our insight into AD. This chapter shows that MEG recordings reflect the alterations on AD patients' brains. These alterations may be associated with deficiencies in information processing.

Keywords: Signal processing, nonlinear methods, Alzheimer's disease, magneto encephalography

1. Introduction

Magnetoencephalography (MEG) is a non-invasive technique that allows recording the magnetic fields produced by the brain electrical currents. This technique provides an excellent temporal resolution [1]. A good spatial resolution can also be achieved due to the large number of sensors. Moreover, the activity in different parts of the brain can be monitored simultaneously with whole-head equipments, such as the magnetometer used in our study [1]. Although EEG and MEG signals originate by the same synchronous oscillations of pyramidal neurons, they reflect slightly different characteristics of the brain activity. Firstly, EEG is sensitive to all primary currents (both tangential and radial components), whereas MEG detects only its tangential components. Secondly, electrical activity is more affected than magnetic oscillations by skull and extracerebral brain tissues [1, 2]. Moreover, EEG acquisition can be significantly influenced by technical and methodological issues, like the sensor placement and the distance between electrodes. Additionally, MEG provides reference-free recordings. On the other hand, the magnetic signals generated by the human brain are extremely weak. Thus, MEG signals can be most conveniently detected by large arrays of superconducting quantum interference devices (SQUIDs) immersed in liquid helium at 4.2 K. In addition, MEG signals must be recorded in a magnetically shielded room to reduce the environmental noise [1]. Therefore, MEG is characterized by high equipment cost and limited availability.

Alzheimer's disease (AD) is a progressive and degenerative brain disorder of unknown aetiology. It is the main cause of dementia in western countries, accounting for 50-60% of all cases [3]. AD affects 1% of population aged 60-64 years, but the prevalence increases exponentially with age, so around 30% of people over 85 years suffer from this disease [4]. Due to the fact that life expectancy has significantly improved in western countries in the last decades, it is expected that the number of people with dementia increase up to 81 million in 2040 [4]. AD is characterized by the presence of neuritic plaques and neurofibrillary tangles, accompanied by the loss of cortical neurons and synapses [5]. Clinically, this disease manifests as a slowly progressive impairment of mental functions whose course lasts several years prior to death, seriously affecting the patient's ability to carry out daily activities. There is no straightforward test for AD diagnosis and a differential diagnosis with other types of dementia and with major depression is usually made. The differential diagnosis includes medical history studies, physical and neurological evaluation, mental status tests, and neuroimaging techniques. Although no treatment can stop the disease, some drugs may help keep symptoms from getting worse for a limited time. Therefore, an early diagnosis of AD is critical to alleviate some symptoms and reduce the damage suffered by the patient's brain.

The electromagnetic brain activity in AD has been researched in the last decades by means of several nonlinear methods. Measures derived from Chaos Theory, as correlation dimension (*D*2) and the first positive Lyapunov exponent (*L*1), have been widely used to study the brain dynamics in AD. *D*2 computes the geometric complexity of the reconstructed attractor [6], whereas *L*1 is a dynamic complexity measure that describes the divergence of trajectories starting at nearby initial states [7]. For instance, Jeong *et al.* [7] showed that AD patients exhibit significantly lower *D*2 and *L*1 values than controls in many EEG channels. These results were confirmed by Jelles *et al.* [6]. In a MEG study [8], this complexity loss was reported only in the high frequency bands. Nevertheless, these classical measures for estimating the nonlinear dynamic complexity have some drawbacks. Reliable estimation of *L*1 and *D*2 requires a large quantity of data and stationary and noise free time series [9], [10]. Since these assumptions cannot be achieved for physiological data, other measures are necessary for the analysis of brain time series.

Several nonlinear measures have been proposed to study the EEG/MEG activity in AD (for review, see [5], [11], [12]). The dimensional complexity of a signal can be estimated directly in the time domain using the fractal dimension (*FD*) without reconstructing the attractor in the phase space. Some *FD* algorithms have been successfully applied to EEG/MEG recordings in dementia [13], [14]. Other nonlinear complexity measures, as Lempel-Ziv complexity (*LZC*) or multiscale entropy (*MSE*), have also been used to characterize the brain activity in AD [15] – [18]. The main result of previous studies is that EEG/MEG activity is less complex in AD patients than in controls.

Entropy is a concept addressing randomness and predictability, with greater entropy often associated with more randomness and less system order [19]. Several entropy measures have demonstrated their usefulness in the analysis of EEG/MEG background activity in AD [17]-[21]. An increase of entropy values has been found using approximate entropy (*ApEn*) [18], [19], sample entropy (*SampEn*) [17], [20], Shannon spectral entropy, Rényi spectral entropy and Tsallis spectral entropy [21]. Therefore, we can conclude that AD is characterized by an increase of regularity of brain activity.

Other nonlinear methods, as detrended fluctuation analysis (*DFA*) and detrended moving average analysis (*DMA*), have been used to detect long-range correlations and fluctuations in EEG and MEG time series [22] – [24]. Abásolo *et al.* [22] applied *DFA* to spontaneous EEG data, finding significant differences between AD patients and controls. Using MEG data, both *DFA* and *DMA* revealed significant differences between both groups [23]. Finally, Stam *et al.* [24] analyzed the fluctuations of the EEG synchronization level at different frequency bands by means of *DFA*.

In this chapter, we have examined the MEG background activity in 36 patients with probable AD and 26 elderly control subjects using three nonlinear measures: *LZC*, *SampEn*, and *DFA*. Our purpose is to test the hypothesis that the neuronal dysfunction in AD is associated with differences in the dynamical processes underlying the MEG recording.

2. Materials

2.1. Subjects

For this chapter, MEG signals were recorded from 62 subjects: 36 AD patients and 26 elderly control subjects. Clinical diagnosis was ascertained by means of exhaustive general medical, neurological, and psychiatric examinations. All patients and controls underwent a neuropsychological evaluation including the Spanish versions of the following scales and batteries: Wechsler Memory Scale 3rd Edition (WMS-III), Boston Naming Test (BNT), Stroop Test, Wisconsin Card Shorting Test (WCST), Silhouettes Test of the Visual Object and Space Battery (VOSP), and tests for constructive and ideatory apraxia. In addition to this, cognitive status was screened in both groups with the Spanish version [25] of the Mini Mental State Examination (MMSE) of Folstein *et al.* [26], whereas functional status was evaluated by means of the Global Deterioration Scale/Functional Assessment Staging (GDS/FAST) system [27].

MEGs were obtained from 36 patients (12 men and 24 women; age = 74.06 ± 6.95 years, mean ± standard deviation, SD) fulfilling the criteria of probable AD. They were recruited from the "Asociación de Familiares de Enfermos de Alzheimer" in Spain. Diagnosis for all patients was made according to the NINCDS–ADRDA (National Institute of Neurological and Communicative Disorders and Stroke and Alzheimer's Disease and Related Disorders Association) criteria [28]. The MMSE and GDS/FAST scores for these patients were 18.06 ± 3.36 and 4.17 ± 0.45 (mean ± SD), respectively. Patients were free of other significant medical, neurological and psychiatric diseases than AD and they were not taking drugs which could affect MEG activity.

The control group consisted of 26 elderly control subjects without past or present neurological disorders (9 men and 17 women; age = 71.77 ± 6.38 years, MMSE score = 28.88 ± 1.18 points, GDA/FAST score = 1.73 ± 0.45 points; mean ± SD). The difference in age between both populations was not statistically significant (p-value = 0.1911 > 0.05). All control subjects and patients' caregivers signed an informed consent for the participation in this research work. The local Ethics Committee approved this study.

2.2. MEG Recording

MEGs were acquired with a 148-channel whole-head magnetometer (MAGNES 2500 WH, 4D Neuroimaging) placed in a magnetically shielded room at "Centro de Magnetoencefalografía Dr. Pérez-Modrego" (Spain). The subjects lay on a patient bed, in a relaxed state and with their eyes closed. For each subject, five minutes of recording were acquired at a sampling frequency of 678.17 Hz, using a hardware band-pass filter from 0.1 to 200 Hz. Then, the equipment decimated each 5 minutes data set. This process consisted of filtering the data to satisfy the Nyquist criterion, following by a down-sampling by a factor of 4, thus obtaining a sampling rate of 169.549 Hz. Finally, artifact-free epochs were processed using a band-pass filter with a Hamming window and cut-off frequencies at 0.5 and 40 Hz. For *LZC*, the epoch length was 20 seconds, as previous studies suggested that *LZC* values become stable for MEG signals longer than 3000 samples [17]. On the other hand, *SampEn* and *DFA* were applied to epochs of 5 seconds (848 samples).

3. METHODOLOGY

3.1. Lempel-Ziv Complexity (LZC)

The *LZC* algorithm was proposed by Lempel and Ziv to evaluate the randomness of finite sequences [29]. It is a nonparametric and simple-to-compute measure of complexity for one-dimensional signals that does not require long data segments to be calculated [30]. Larger *LZC* values correspond to more complex data. Due to the fact that *LZC* analyzes a finite symbol sequence, the given signal must first be coarse-grained [30]. In this study, a binary (zeros and ones) conversion was used, since previous studies found that this kind of conversion may keep enough signal information [30]. The data points were compared to a threshold, T_d. We fixed T_d to the median of the analyzed signal, as partitioning about this value is robust to outliers [31]. By comparison with T_d, the original data $X = (x_1, x_2,..., x_N)$ are converted into a 0-1 sequence $P = (s_1, s_2,..., s_N)$, with s_i defined by [30]:

$$s_i = \begin{cases} 0 & if\ x_i < T_d \\ 1 & if\ x_i \geq T_d \end{cases} \tag{1}$$

The string *P* is scanned from left to right and a complexity counter *c*(*N*) is increased by one unit every time a new subsequence of consecutive characters is encountered in the scanning process. The detailed algorithm for the measure of *LZC* is as follows [30] – [32]:

1) Let *S* and *Q* denote two subsequences of the original sequence *P*. *SQ* is the concatenation of *S* and *Q*, while $SQ\pi$ is a string derived from *SQ* after its last character is deleted (π means the operation to delete the last character). Let $v(SQ\pi)$ denote the vocabulary of all different substrings of $SQ\pi$.
2) At the beginning, the complexity counter $c(N) = 1$, $S = s_1$, $Q = s_2$, $SQ = s_1, s_2$ and $SQ\pi = s_1$.
3) For generalization, suppose that $S = s_1, s_2,\ldots, s_r$, $Q = s_{r+1}$ and, therefore, $SQ\pi = s_1, s_2,\ldots, s_r$. If $Q \in v(SQ\pi)$, then *Q* is a subsequence of $SQ\pi$, not a new sequence.
4) *S* does not change and renew *Q* to be s_{r+1}, s_{r+2}, then judge if *Q* belongs to $v(SQ\pi)$ or not.
5) The previous steps are repeated until *Q* does not belong to $v(SQ\pi)$. Now $Q = s_{r+1}, s_{r+2},\ldots, s_{r+i}$ is not a subsequence of $SQ\pi = s_1, s_2,\ldots, s_{r+i-1}$, so increase the counter by one.
6) Thereafter, *S* and *Q* are combined and renewed to be $s_1, s_2,\ldots, s_r, s_{r+1},\ldots, s_{r+i}$, and s_{r+i+1}, respectively.
7) Repeat the previous steps until *Q* is the last character. At this time, the number of different substrings is *c*(*N*), the measure of complexity.

In order to obtain a complexity measure independent of the sequence length, $c(N)$ should be normalized. If the length of the sequence is N and α is the number of different symbols, it has been proved that the upper bound of $c(N)$ is given by [29]:

$$c(N) < \frac{N}{(1-\varepsilon_N)\log_\alpha(N)} \tag{2}$$

where ε_N is a small quantity and $\varepsilon_N \to 0$ ($N \to \infty$). In general, $N/\log_\alpha(N)$ is the upper limit of $c(N)$, i.e.

$$\lim_{N\to\infty} c(N) = b(N) \equiv \frac{N}{\log_\alpha(N)} \tag{3}$$

For a binary conversion $\alpha = 2$, $b(N) \equiv N/\log_2(N)$ and $c(N)$ can be normalized via $b(N)$:

$$C(N) = \frac{c(N)}{b(N)} \tag{4}$$

$C(N)$ reflects the arising rate of new patterns along with the sequence.

3.2. Sample Entropy (Sampen)

Sample entropy (*SampEn*) is an embedding entropy that quantifies the signal irregularity: more irregularity in the data produces larger *SampEn* values [33]. This metric solves some problems associated with *ApEn*. The *ApEn* algorithm counts each sequence as matching itself to avoid the occurrence of ln(0) in the calculations and this has led to discussion of the bias of *ApEn* [33]. To reduce this bias, Richman and Moorman have developed the so called *SampEn*. *SampEn* is largely independent of the signal length and displays relative consistency under circumstances where *ApEn* does not. Additionally, the algorithm used to compute the *SampEn* is simpler than the *ApEn* one [33]. *SampEn* has two input parameters: a run length m and a tolerance window r. *SampEn* is the negative natural logarithm of the conditional probability that two sequences similar for m points remain similar at the next point [33]. For this study, *SampEn* was calculated with parameter values $m = 1$ and $r = 0.25$ times the SD of the original data sequence.

To calculate the $SampEn(m, r, N)$ from a time series, $X = (x_1, x_2,..., x_N)$, one should follow these steps [33]:

1) Form a set of vectors $X_m^1,\ldots, X_m^{N-m+1}$ defined by $X_m^i = (x_i, x_{i+1},\ldots, x_{i+m-1})$, $i = 1,\ldots, N - m + 1$.
2) The distance between X_m^i and X_m^j, $d[X_m^i, X_m^j]$, is the maximum absolute difference between their respective scalar components:

$$d[X_m^i, X_m^j] = \max_{k=0,\ldots m-1} | x_{i+k} - x_{j+k} | \tag{5}$$

3) For a given $X_m^{\ i}$, count the number of j ($1 \leq j \leq N - m$, $j \neq i$), denoted as B_i, such that $d[X_m^{\ i}, X_m^{\ j}] \leq r$. Then, for $1 \leq i \leq N - m$,

$$B_i^m(r) = \frac{1}{N-m-1} B_i \tag{6}$$

4) Define as:

$$B^m(r) = \frac{1}{N-m} \sum_{i=1}^{N-m} B_i^m(r) \tag{7}$$

5) Similarly, calculate as $1/(N - m + 1)$ times the number of j ($1 \leq j \leq N - m$, $j \neq i$), such that the distance between $X_{m+1}^{\ j}$ and $X_{m+1}^{\ i}$ is less than or equal to r. Set $A^m(r)$ as:

$$A^m(r) = \frac{1}{N-m} \sum_{i=1}^{N-m} A_i^m(r) \tag{8}$$

Thus, $B^m(r)$ is the probability that two sequences will match for m points, whereas is the probability that two sequences will match for $m + 1$ points.

6) Finally, define:

$$SampEn(m,r) = \lim_{N \to \infty} \left\{ -\ln \left[\frac{A^m(r)}{B^m(r)} \right] \right\} \tag{9}$$

which is estimated by the statistic:

$$SampEn(m,r,N) = -\ln \left[\frac{A^m(r)}{B^m(r)} \right] \tag{10}$$

3.3. Detrended Fluctuation Analysis (DFA)

DFA is a measure widely used for the detection of long-range correlations and fluctuations in time series [34]. This method provides a simple quantitative parameter (the scaling exponent) to represent the correlation properties of a signal [35]. *DFA* permits the detection of long-range correlations embedded in seemingly non-stationary time series. Additionally, it avoids the spurious detection of apparent long-range correlations that are an artifact of non-stationarity [35]. Given a one dimensional time series $X = (x_1, x_2, \ldots, x_N)$, we describe the algorithm to compute the *DFA* [34], [36]:

1) Integrate the signal X:

$$y_i = \sum_{j=1}^{i} \left[x_j - \bar{X} \right] \tag{11}$$

where represents the mean value of *X*.

2) The integrated time series *Y* is divided into windows of equal size *k*. The size of the windows is ranged between 3 and 84, as one-tenth of the signal length can be considered as the maximum window size when using *DFA* [35].
3) Within each window, labeled *b* (*b* = 1, 2,..., *B*), perform a least-square fit of *Y* by a straight line Y^b. This is the semilocal trend for the bth window.
4) Define to be the variance of the fluctuation *Y* from in the bth window:

$$F_b^2(k) = \frac{1}{k} \sum_{t=(b-1)k+1}^{bk} \left(y_t - y_t^b\right)^2 \tag{12}$$

5) The square root of the average of over all windows is the rms fluctuation from the semilocal trends in *B* windows, each having *k* time points:

$$F(k) = \sqrt{\frac{1}{B} \sum_{b=1}^{B} F_b^2(k)} \tag{13}$$

6) Finally, the study of the dependence of *F*(*k*) on the window size *k* is the essence of *DFA*. If it is a power-law behavior $F(k) \propto k^{\alpha}$, the scaling exponent is an indicator of the nature of the fluctuations in the MEG signals.

4. Results

LZC, *SampEn*, and *DFA* were estimated for the 148 MEG channels. The results were averaged based on all the artefact-free epochs within the five-minute period of recording. Additionally, to simplify the statistical analyses, the results were averaged over all channels. Normality of distribution was assessed with Kolmogorov-Smirnov test, whereas homoscedasticity was analyzed with Levene's test. As the nonlinear results did not meet homoscedasticity assumption, Welch *t*-test was used for the statistical comparison between AD patients and control subjects.

LZC is a non-parametric method which measures the number of different substrings and the rate of their recurrence along the original time series [29]. In this study, a binary conversion was used for the coarse-grained process. Then, the number of different subsequences in the original binary sequence is counted. For this measure, 20 s artefact-free epochs (3392 samples) were employed due to the fact that a previous study showed that the *LZC* values become stable for MEG signals longer than 3000 samples [16]. The average *LZC* values were 0.58 ± 0.07 (mean ± SD) for AD patients and 0.65 ± 0.08 for controls. Figure 1 shows the mean values obtained for each group. For statistical comparison, Welch *t*-test was applied to the mean *LZC* values, averaged over all channels. A statistically significant *p*-value of 0.0013 was obtained.

SampEn algorithm was applied to all 148 MEG channels to measure the signal regularity. Although both *m* and *r* are critical parameters in the performance of *SampEn*, there are no guidelines for optimizing their values. In this study, we have chosen $m = 1$ and $r = 0.25$ times the SD of the original time series. The average *SampEn* values for AD and control groups were 1.05 ± 0.21 and 1.19 ± 0.20, respectively. Our results showed that *SampEn* values were higher in the control group than in the AD group for all channels (see Figure 2), which suggests that AD is accompanied by a MEG regularity increase. Additionally, the differences between both groups were statistically significant: p-value = 0.0099 (Welch t-test).

MEG data were also analyzed with *DFA* method, using window sizes between 3 and 84 samples. Our results showed two scaling regions with different slopes. We have denoted α_1 the scaling exponent of the first region ($3 \leq k \leq 7$), and α_2 the exponent of the second one ($14 \leq k \leq 73$). Mean values of the scaling exponents were the following: 1.70 ± 0.06 (α_1 for AD patients), 1.71 ± 0.06 (α_1 for controls), 0.23 ± 0.09 (α_2 for AD), and 0.15 ± 0.07 (α_2 for control group). No significant differences were found between the α_1 values of AD patients and control subjects: p-value = 0.6214 (Welch t-test). On the other hand, the differences between both groups were statistically significant when the slopes of the second scaling region were analyzed (p-value = 0.0003). Figure 3 illustrates the mean α_2 values for AD and control groups.

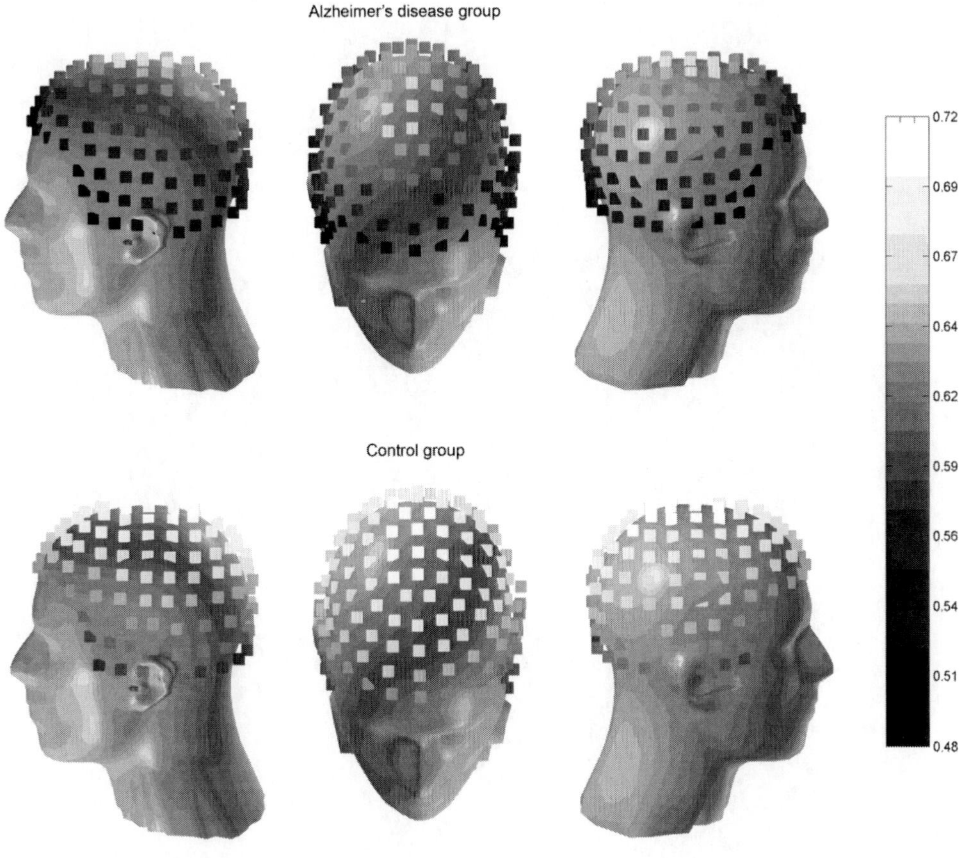

Figure 1. Average *LZC* values from MEGs in AD patients and control subjects for all channels.

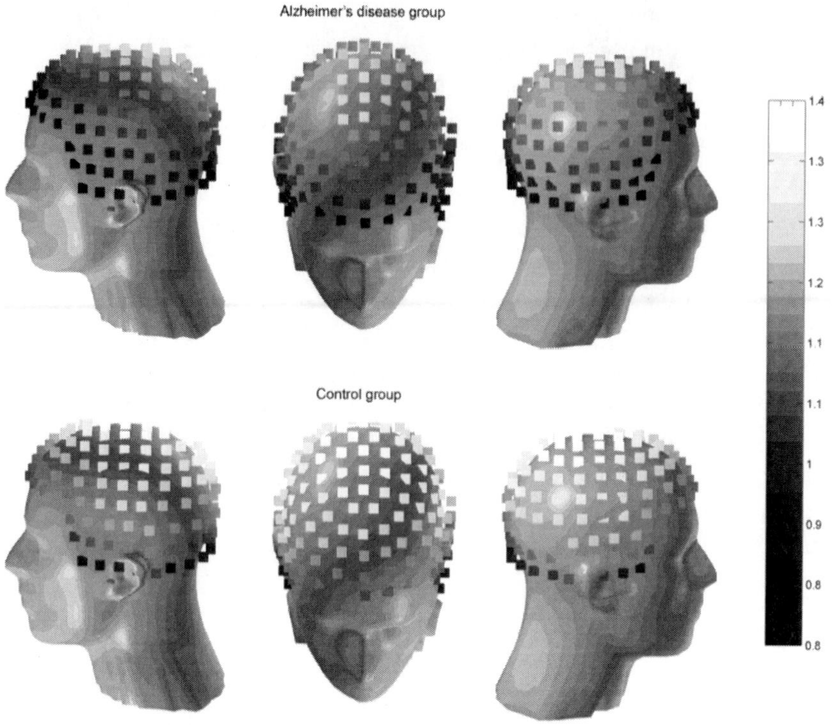

Figure 2. Average *SampEn* values from MEGs in AD patients and control subjects for all channels.

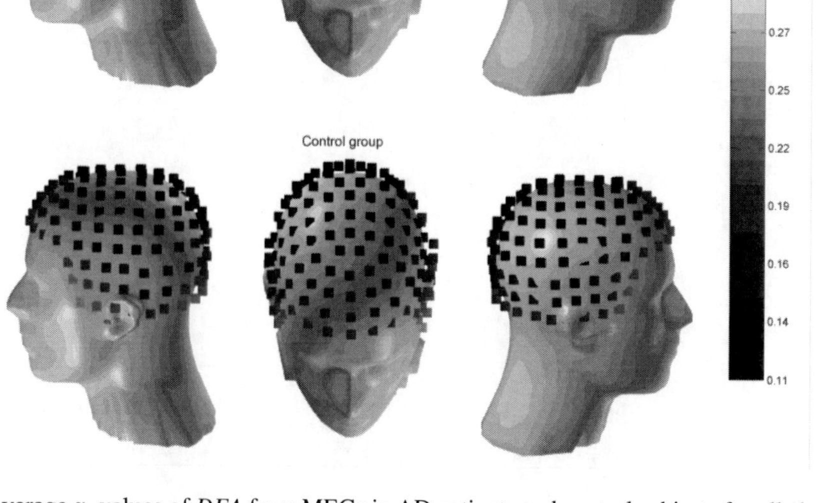

Figure 3. Average α_2 values of *DFA* from MEGs in AD patients and control subjects for all channels.

Table 1. Sensitivity, specificity, accuracy, and AUC values obtained with each nonlinear method

Method	Sensitivity	Specificity	Accuracy	AUC
LZC	0.8056	0.6154	0.7258	0.7863
SampEn	0.7778	0.5000	0.6613	0.7126
DFA (α_2)	0.7500	0.6154	0.6935	0.7959

Finally, receiver operating characteristic (ROC) curves with a leave-one-out cross-validation procedure were used to assess the ability of the three methods to classify AD patients and control subjects. A ROC curve summarizes the performance of a two-class classifier across the range of possible thresholds. It is a graphical representation of the trade-offs between sensitivity and specificity. Sensitivity is the true positive rate whereas specificity is equal to the true negative rate. Accuracy is the percentage of subjects (AD patients and controls) correctly recognized. Finally, the area under the ROC curve (AUC) is a measure of the classifier performance. To simplify the analyses, the results of each method were averaged over all channels. The highest accuracy (72.58%) and AUC (0.7959) values were reached using *LZC* and *DFA*, respectively. Sensitivity, specificity, accuracy, and AUC values obtained with each measure are displayed in Table 1.

5. Discussion

We have analyzed the spontaneous MEG activity from 36 patients with probable AD and 26 elderly control subjects by means of three nonlinear methods: *LZC*, *SampEn* and *DFA*. Our purpose was to check the hypothesis that MEG background activity is different in AD patients and control subjects.

LZC has proven to be effective in discriminating AD patients from controls subjects. Our study revealed that AD patients have lower *LZC* values than controls at all channels. These results are in agreement with previous research works that have applied nonlinear methods to estimate the complexity of the AD patients' brain activity. During the last two decades, the nonlinear complexity measure most used to study the brain activity in AD is *D*2, often accompanied by *L*1 [6] – [8], [37] – [41]. *D*2 is considered to be a reflection of the complexity of the cortical dynamics underlying bran recordings. Thus, reduced *D*2 values of the EEG/MEG in AD patients indicate that brains injured by AD exhibit a decrease in complexity of the electromagnetic brain activity [5]. Besthorn *et al.* [41] suggested that this reduced complexity in AD may be associated with an increase in the proportion of lower-frequency component in the patients' EEGs. The so-called *L*1 describes the divergence of trajectories starting at nearby initial states [7]. Previous studies showed that AD patients exhibit lower *L*1 values than control subjects [7], [39]. This decrease would indicate a drop in the flexibility of the brain to process the information [7]. Recent studies have confirmed this complexity decrease in AD with different kinds of complexity measures. For instance, Escudero *et al.* [15] found significant differences in some EEG channels with *MSE*. Other EEG/MEG studies demonstrated that AD patients had lower *LZC* values than controls [16], [17], [42]. In a subsequent MEG study [13], this complexity loss in AD was confirmed using

HFD. Finally, Henderson *et al.* [14] suggested that the zero-set *FD* of the EEG might be used to detect AD within the general at risk population (specificity 99.9%, sensitivity 67%).

Our results also showed that AD patients have lower *SampEn* values than controls. Moreover, differences between both groups were statistically significant. Our results are in agreement with previous research works that have applied different entropy measures to estimate the regularity of EEG/MEG recordings from AD patients. In EEG, *ApEn* values were significantly lower in AD patients at parietal electrodes [19]. With *SampEn*, Abásolo *et al.* [20] found significant differences at P3, P4, O1 and O2. These results were subsequently confirmed by MEG studies [17], [18]. Other studies have shown an increase of the EEG/MEG regularity in AD patients compared with control subjects using different kinds of spectral entropies [20], [21].

DFA results revealed two scaling regions in MEGs from both AD patients and controls. The existence of two scaling regions had been previously suggested in [22], [23]. Significant differences between groups were found at the second scaling exponent, but not at the first one. These results agree with a previous EEG study [22]. Other EEG research revealed that spontaneous fluctuations of synchronization were diminished in AD patients in the lower alpha and beta bands [24]. In a previous MEG study, *DFA* and also *DMA* showed the fluctuation changes in AD patients' MEG signals [23].

ROC curves with a leave-one-out cross-validation procedure were used to assess the ability of *LZC*, *SampEn* and *DFA* to classify AD patients and control subjects. Using *LZC*, an accuracy of 72.58% (80.56%, sensitivity; 61.54% specificity) was achieved. With *SampEn*, accuracy and AUC values of 66.13% and 0.7126 were obtained. Finally, specificity of 61.54%, sensitivity of 75.00%, accuracy of 69.35%, and AUC of 0.7959 were reached when the second scaling exponent of *DFA* was analyzed. In previous papers, these and other nonlinear methods have been used to distinguish AD patients and control subjects. For instance, an accuracy of 70.7% was obtained with *LZC* in a MEG study [18]. In the same research, the accuracy values achieved with *SSE* and *ApEn* were 70.7% and 51.2%, respectively [18]. When the auto-mutual information decrease rate was applied to MEG data, an accuracy of 82.9% was reached [43]. The highest accuracy in MEG complexity studies was achieved when mean *HFD* values were analyzed with a ROC curve (87.8%) [13]. This value was also obtained using spectral entropies [21]. The accuracies achieved in EEG studies on AD with nonlinear measures are the following: 81.8% using *LZC* [42], 77.3% with *SampEn* [20], 90.1% with *MSE* [15] and 69.5% using the classical measure *D*2 [38]. Nevertheless, all these values, both in EEG and MEG research works, should be taken with caution due to the small sample sizes.

CONCLUSION

In sum, this chapter leads us to conclude that neuronal dysfunction in AD is associated with differences in the dynamical processes underlying the spontaneous MEG activity. The results obtained with *LZC*, *SampEn* and *DFA* showed significant differences between AD patients and controls. Although these changes seem to be associated with the deficiencies in information processing suffered by AD patients, its pathophysiological implications are not clear. Among others, three mechanisms can be responsible for it: neuronal death, a general effect of neurotransmitter deficiency, and connectivity loss of local neural networks [5].

Our results indicate that nonlinear measures could be useful in AD diagnosis. Nevertheless, some limitations of our study merit consideration. Firstly, the sample size is small to prove the usefulness of these measures as diagnostic tools. Moreover, the detected changes in spontaneous MEG activity are not specific to AD, appearing in other brain disorders. Finally, the results obtained from each parameter were averaged to simplify the analyses, loosing the spatial information of the MEG signals. Future efforts will be focussed to increase the MEG database, as well as to extend the results to other neurodegenerative diseases.

ACKNOWLEDGMENTS

This research was supported in part by Ministerio de Economía y Competitividad and FEDER under project TEC2011-22987, the Proyecto Cero 2011 on Ageing from Fundación General CSIC, Obra Social La Caixa and CSIC and project VA111A11-2 from Consejería de Educación (Junta de Castilla y León).

REFERENCES

[1] R. Hari, "Magnetoencephalography in clinical neurophysiological assessment of human cortical functions," in *Electroencephalography: Basic Principles, Clinical Applications, and Related Fields*, 5th ed., E. Niedermeyer and F. Lopes da Silva, Eds. Philadelphia: Lippincontt Williams and Wilkins, 2005, pp. 1165–1197.

[2] M. Hämäläinen, R. Hari, R. Ilmoniemi, J. Knuutila, and O.V. Lounasmaa, "Magnetoencephalography: theory, instrumentation, and applications to noninvasive studies of the working human brain," *Rev. Mod. Phys.*, vol. 65, pp. 413–97, 1993.

[3] K. Blennow, M. J. de Leon, and H. Zetterberg, "Alzheimer's disease," *Lancet*, vol. 368, pp. 387–403, 2006.

[4] F. Jorm, "Cross-national comparisons of the occurrence of Alzheimer's and vascular dementias," *Eur. Arch. Psychiatry Clin. Neurosci.*, vol. 240, pp. 218–222, 1991.

[5] J. Jeong, "EEG dynamics in patients with Alzheimer's disease," *Clin. Neurophysiol.*, vol. 115, pp. 1490–1505, 2004.

[6] B. Jelles, J. H. van Birgelen, J. P. J. Slaets, R. E. M. Hekster, E. J. Jonkman, and C. J. Stam, "Decrease of non-linear structure in the EEG of Alzheimer patients compared to healthy controls," *Clin. Neurophysiol.*, vol. 110, pp. 1159–1167, 1999.

[7] J. Jeong, S. Y. Kim, and S. H. Han, "Non-linear dynamical analysis of the EEG in Alzheimer's disease with optimal embedding dimension," *Electroenceph. Clin. Neurophysiol.*, vol. 106, pp. 220–228, 1998.

[8] M. van Cappellen van Walsum, Y. A. L. Pijnenburg, H. W. Berendse, B. W. van Dijk, D. L. Knol, Ph. Scheltens, and C. J. Stam, "A neural complexity measure applied to MEG data in Alzheimer's disease," *Clin. Neurophysiol.*, vol. 114, pp. 1034–1040, 2003.

[9] J. P. Eckmann and D. Ruelle, "Fundamental limitations for estimating dimensions and Lyapunov exponents in dynamical systems," *Physica D*, vol. 56, pp. 185–187, 1992.

[10] P. Grassberger and I. Procaccia, "Characterization of strange attractors," *Phys. Rev. Lett.*, vol. 50, pp. 346–349, 1983.

[11] C. J. Stam, "Nonlinear dynamical analysis of EEG and MEG: Review of an emerging field," *Clin. Neurophysiol.*, vol. 116, pp. 2266–2301, 2005.

[12] C. J. Stam, "Use of magnetoencephalography (MEG) to study functional brain networks in neurodegenerative disorders," *J. Neurol. Sci.*, vol. 289, pp. 128–134, 2010.

[13] C. Gómez, A. Mediavilla, R. Hornero, D. Abásolo, and A. Fernández, "Use of the Higuchi's fractal dimension for the analysis of MEG recordings from Alzheimer's disease patients," *Med. Eng. Phys.*, vol. 31, pp. 306–313, 2009.

[14] G. Henderson, E. Ifeachor, N. Hudson, C. Goh, N. Outram, S. Wimalartna, C. Del Percio, and F. Vecchio, "Development and assessment of methods for detecting dementia using the human electroencephalogram," *IEEE Trans. Biomed. Eng.*, vol. 53, pp. 1557–1568, 2006.

[15] J. Escudero, D. Abásolo, R. Hornero, P. Espino, and M. López, "Analysis of electroencephalograms in Alzheimer's disease patients with multiscale entropy," *Physiol. Meas.*, vol. 27, pp. 1091–1106, 2006.

[16] C. Gómez, R. Hornero, D. Abásolo, A. Fernández, and M. López, "Complexity analysis of the magnetoencephalogram background activity in Alzheimer's disease patients," *Med. Eng. Phys.*, vol. 28, pp. 851–859, 2006.

[17] C. Gómez, R. Hornero, D. Abásolo, A. Fernández, and J. Escudero, "Analysis of MEG background activity in Alzheimer's disease using non-linear methods and ANFIS," *Ann. Biomed. Eng.*, vol. 37, pp. 586–594, 2009.

[18] R. Hornero, J. Escudero, A. Fernández, J. Poza, and C. Gómez, "Spectral and non-linear analyses of MEG background activity in patients with Alzheimer's disease," *IEEE Trans. Biomed. Eng.*, vol. 55, pp. 1658–1665, 2008.

[19] D. Abásolo, R. Hornero, P. Espino, J. Poza, C. I. Sánchez, and R. de la Rosa, "Analysis of regularity in the EEG background activity of Alzheimer's disease patients with approximate entropy," *Clin. Neurophysiol.*, vol. 116, pp. 1826–1834, 2005.

[20] D. Abásolo, R. Hornero, P. Espino, D. Álvarez, and J. Poza, "Entropy analysis of the EEG background activity in Alzheimer's disease patients," *Physiol. Meas.*, vol. 27, pp 241–253, 2006.

[21] J. Poza, R. Hornero, J. Escudero, A. Fernández, and C. I. Sánchez, "Regional analysis of spontaneous MEG rhythms in patients with Alzheimer's disease using spectral entropies," *Ann. Biomed. Eng.*, vol. 36, pp. 141–152, 2008.

[22] D. Abásolo, R. Hornero, J. Escudero, and P. Espino, "A study of the possible usefulness of detrended fluctuation analysis of the electroencephalogram background activity in Alzheimer's disease," *IEEE Trans. Biomed. Eng.*, vol. 55, pp. 2171–2179, 2008.

[23] C. Gómez, R. Hornero, D. Abásolo, A. Fernández, and J. Poza, "Study of the MEG background activity in Alzheimer's disease patients with scaling analysis methods," in: *31st Annual International Conference of the IEEE Engineering in Medicine and Biology Society*, pp. 3485–3488, 2009.

[24] C. J. Stam, T. Montez, B. F. Jones, S. A. R. B. Rombouts, Y. van der Made, Y. A. L. Pijnenburg, and Ph. Scheltens, "Disturbed fluctuations of resting state EEG

synchronization in Alzheimer's disease," *Clin. Neurophysiol.*, vol. 116, pp. 708–715, 2005.

[25] Lobo, J. Ezquerra, F. Gómez Burgada, J. M. Sala, and A. Seva Díaz, "Cognocitive mini-test (a simple practical test to detect intellectual changes in medical patients)," *Actas Luso-Españolas de Neurología, Psiquiatría y Ciencias Afines*, vol. 7, pp. 189–202, 1979.

[26] M. F. Folstein, S. E. Folstein, and P. R. McHugh, "Mini-mental state. A practical method for grading the cognitive state of patients for the clinician," *J. Psychiatr. Res.*, vol. 12, pp. 189–198, 1975.

[27] S. Auer and B. Reisberg, "The GDS/FAST staging system," *Int. Psychogeriatr.*, vol. 9, pp. 167–171, 1997.

[28] G. McKhann, D. Drachman, M. Folstein, R. Katzman, D. Price, and E. M. Stadlan, "Clinical diagnosis of Alzheimer's disease: report of the NINCDS-ADRDA Work Group under the auspices of Department of Health and Human Services Task Force on Alzheimer's Disease," *Neurology*, vol. 34, pp. 939–944, 1984.

[29] A. Lempel and J. Ziv, "On the complexity of finite sequences," *IEEE Trans. Inf. Theory*, vol. IT-22, pp. 75–81, 1976.

[30] X. S. Zhang, R. J. Roy, and E. W. Jensen, "EEG complexity as a measure of depth of anesthesia for patients," *IEEE Trans. Biomed. Eng.*, vol. 48, pp. 1424–1433, 2001.

[31] R. Nagarajan, "Quantifying physiological data with Lempel-Ziv complexity - certain issues," *IEEE Trans. Biomed. Eng.*, vol. 49, pp. 1371–1373, 2002.

[32] X. S. Zhang and R. J. Roy, "Derived fuzzy knowledge model for estimating the depth of anesthesia," *IEEE Trans. Biomed. Eng.*, vol. 48, pp 312–323, 2001.

[33] J. S. Richmann and J. R. Moorman, "Physiological time-series analysis using approximate entropy and sample entropy," *Am. J. Physiol. Heart Circ. Physiol.*, vol. 278, pp. H2039–H2049, 2000.

[34] C. K. Peng, S. V. Buldyrev, S. Havlin, M. Simons, H. E. Stanley, and A. L Goldberger, "Mosaic organization of DNA nucleotides," *Phys. Rev. E*, vol. 49, pp. 1685–1689, 1994.

[35] K. Hu, P. C. Ivanov, Z. Chen, P. Carpena, and H. E. Stanley, "Effects of trends on detrended fluctuation analysis," *Phys. Rev. E*, vol. 64, pp. 011114, 2001.

[36] R. C. Hwa and T. C. Ferre, "Scaling properties of fluctuations in the human electroencephalogram," *Phys. Rev. E*, vol. 66, pp. 021901, 2002.

[37] W. S. Pritchard, D. W. Duke, K. L. Coburn, N. C. Moore, K. A. Tucker, M. W. Jann, and R. M. Hostetler, "EEG-based, neural-net predictive classification of Alzheimer's disease versus control subjects is augmented by nonlinear EEG measures," *Electroencephalogr. Clin. Neurophysiol.*, vol. 91, pp. 118–130, 1994.

[38] C. Besthorn, R. Zerfass, C. Geiger-Kabisch, H. Sattel, S. Daniel, U. Schreiter-Gasser, and H. Förstl, "Discrimination of Alzheimer's disease and normal aging by EEG data," *Electroenceph. Clin. Neurophysiol*, vol. 103, pp. 241–248, 1997.

[39] J. Jeong, J. H. Chae, S. Y. Kim, and S. H. Han, "Nonlinear dynamic analysis of the EEG in patients with Alzheimer's disease and vascular dementia," *J. Clin. Neurophysiol.*, vol. 18, pp. 58–67, 2001.

[40] C. J. Stam, B. Jelles, H. A. M. Achtereekte, S. A. R. B. Rombouts, J. P. J. Slaets, and R. W. M. Keunen, "Investigation of EEG nonlinearity in dementia and Parkinson's disease," *Electroenceph. Clin. Neurophysiol.*, vol. 95, pp. 309–317, 1995.

[41] C. Besthorn, H. Sattel, C. Geiger-Kabisch, R. Zerfass, and H, Förstl, "Parameters of EEG dimensional complexity in Alzheimer's disease," *Electroenceph. Clin. Neurophysiol.*, vol. 95, pp. 84–89, 1995.
[42] D. Abásolo, R. Hornero, C. Gómez, M. García, and M. López, "Analysis of EEG background activity in Alzheimer's disease patients with Lempel–Ziv complexity and Central Tendency Measure," *Med. Eng. Phys.*, vol. 28, pp. 315–322, 2006.
[43] C. Gómez, R. Hornero, D. Abásolo, A. Fernández, and J. Escudero, "Analysis of the magnetoencephalogram background activity in Alzheimer's disease patients with auto-mutual information," *Comput. Meth. Programs Biomed.*, vol. 87, pp. 239–247, 2007.

In: Focus on Nonlinear Analysis Research
Editors: Gualtiero Padovani and Manuel Occhino
ISBN: 978-1-62417-350-9
© 2013 Nova Science Publishers, Inc.

Chapter 5

MESHFREE-ENRICHED FINITE ELEMENT METHOD FOR NONLINEAR ANALYSIS OF PATH-INDEPENDENT AND PATH-DEPENDENT PROBLEMS IN SOLIDS

C. T. Wu and W. Hu*
Livermore Software Technology Corporation, Livermore, CA, US

ABSTRACT

This chapter presents a large deformation analysis of nonlinear path-independent and path-dependent problems based on Meshfree-enriched Finite element Method (ME-FEM). In ME-FEM, the problem domain in two-dimension is first discretized by a regular triangulation using linear triangular elements. The element formulation is then established by introducing a meshfree convex approximation into the linear triangular element along with an enriched meshfree node. To provide a volumetric locking-free analysis for near-incompressible materials, an area-weighted smoothing scheme for deformation gradient is developed in conjunction with the meshfree-enriched element interpolation functions to yield a discrete divergence-free property at the integration points for both path-independent and path-dependent problems. A three-field Hu-Washizu variational functional is considered for the development of nonlinear finite element formulation. Since the meshfree-enriched element shape functions are constructed using the meshfree convex approximation, they pose the desired Kronecker-delta property at the element edge thus requires no special treatments in the enforcement of essential boundary condition as well as the contact conditions. The variational formulation is further degenerated to become a primal problem based on the proposed area-weighted smoothing scheme for deformation gradient. As the results, the proposed method can be easily implemented into a conventional displacement-based finite element code. Several hyper-elasticity and elasto-plasticity problems are studied to demonstrate the effectiveness of the proposed method. The numerical results indicated that the proposed method is capable of delivering a volumetric locking-free and pressure oscillation-free solution solutions to near-incompressible materials in large deformation.

---

* Livermore Software Technology Corporation, Livermore, CA 94550, US.

Keywords: Meshfree, finite element, nonlinear, near-incompressible, plasticity

1. Introduction

It is a well-known fact that the standard displacement-based finite element method suffers from severe locking issue in simulating nearly incompressible material because of its over-constrained nature. Elastomers and metals are the typical examples with nearly incompressible material behavior: the characteristics of incompressibility in elastomers is defined by the large ratio of the bulk modulus to the shear modulus; isochoricity of the plastic deformation presents incompressible constraints in metal plasticity. On the other hand, elastomers and metals are highly extensible. For example, the elastomeric components in tire compounds often experience excessive strains in the practice which could be in the order of several hundred percent [43]. The incompressibility constraints and large material deformation have presented tremendous difficulties in the numerical simulation using the standard displacement-based finite element formulation.

The incompressible locking in finite element methods has been studied extensively and many special numerical techniques have been proposed to resolve this difficulty. Among them are reduced/selective integration [24], reduced integration and hourglass control method [8], Taylor expansion method [29], mixed formulation [46], pressure projection method [11, 12] and average nodal pressure element [9]. The approach based on mixed formulation has a close link with Hughes's reduced/selective integration [31] where the displacement and pressure interpolant spaces are subjected to an inf-sup condition [5] for the stability requirement. In Taylor expansion method, an assumed strain field [37] is obtained by the Taylor expansion of displacement gradient matrix and the resulting discrete equation can be expressed explicitly by one-point quadrature terms and their stabilization [29]. In pressure projection method, the pressure computed from the displacement field is projected onto a lower-order space by a least-squares projection at the element level. The resulting equilibrium equation is equivalent to the perturbed Lagrangian formulation [13]. Since the pioneering work by Bonet *et al.* [9] various average nodal pressure formulations [1, 26] have been developed to overcome incompressible locking. A *priori* error estimate [27] using primal and dual meshes reveals that original average nodal pressure formulation does not satisfy a uniform inf-sup condition [5]. In order to have a stable formulation, the linear displacement space needs to be enriched with bubble functions as in the mini-element [2]. This analysis leads to a consistent variational framework for the stabilized nodally integrated tetrahedral presented by Puso *et al.* [35]. Another pressure averaging approach [21, 22] based on macroelement technique [39] also has been developed for near-incompressible elasticity problems leading to the uniform convergence in the nearly incompressible case.

Alternatively several non-traditional numerical methods such as meshfree methods [7, 30] and generalized finite element methods [6, 19] have been proposed to solve the incompressible locking problem. A pseudo-divergence-free interpolation for Element-free Galerkin method [7] was proposed by Vidal et al. [41] to diffuse the divergence-free constraint which can be imposed a *priori* in a displacement-based Galerkin meshfree formulation. Another locking-free displacement-based Galerkin meshfree formulation which is an extension of finite element projection method [11, 12] was presented by Chen *et al.* [14]

for the nonlinear analysis of rubber-like materials. Subsequently, various meshfree approaches have also been developed to alleviate the incompressible locking [16, 17, 32] in the framework of B-bar or mixed formulations. Dolbow and Devan [18] presented a geometrically nonlinear assumed strain method for the nonlinear analysis of hyperelastic materials involving displacement discontinuity where the strain is enriched based on the generalized finite element approach. Similar idea of using generalized finite element approach was presented by Srinivasan *et al.* [38] in the framework of mixed finite element method for nonlinear analysis of near-incompressible rubber compounds. Recently, the iso-geometric discretization based on Non-Uniform Rational B-Splines (NURBS) [25] has presented a promising alternative to solve the incompressible or near-incompressible problems. The high continuity of the NURBS interpolation allows us to solve the incompressible elasticity as an elliptic fourth-order problem in terms of a scalar stream function whose *curl* gives the displacement field [3]. A nonlinear F-bar projection method [20] using the higher-order NURBS interpolation was also proposed for the nonlinear analysis of near-incompressible elasticity. More recently, a meshfree-enriched finite element method (ME-FEM) was proposed by Wu and Hu [45] to overcome the incompressible locking problem. The meshfree-enriched linear element is established by introducing a first-order convex meshfree approximation [33, 43, 44] into a linear element with an enriched meshfree node. Additional strain smoothing procedure [45] is developed in corporation with the meshfree-enriched finite element interpolation to acquire the discrete divergence-free property for the locking-free analysis. An equivalent mixed formulation was also derived in [45] for the stability study of triangular and tetrahedral elements. Their numerical inf-sup [4] study indicates the pair of spaces in displacement and pressure fields is inf-sup stable.

The purpose of this paper is to present a nonlinear version of the meshfree-enriched finite element formulation using triangular elements with a smoothed deformation gradient for general large deformation analysis. The reminder of the paper is outlined as follows: In the next Section, we provide an overview on the meshfree-enriched triangular element. In addition, a smoothing procedure on deformation gradients is introduced. In Section 3, we present a total Lagrangian formulation and updated Lagrangian formulation of the meshfree-enriched finite element method for the nonlinear hyperelasticity and elaso-plasticity, respectively. A mixed three-field Hu-Washizu-de Veubeke variational principle is utilized to formulate the problem. An equivalent assumed strain method is described and the discrete equation is derived. Several numerical examples are presented in Section 4 to demonstrate the accuracy and robustness of the proposed method in the two-dimensional large deformation analysis. Final remarks are given in section 5.

2. Element Formulation

2.1. Meshfree-Enriched Finite Element Triangular Elements

This section provides an overview on the construction of meshfree-enriched finite element interpolations in triangular elements. Let's consider a locally quasi-uniform triangulation of the polygonal domain Ω, where consists of simplexes and is denoted by $\mathbf{M}_h = \cup_e \mathrm{T}_e$. Each simplex or triangle contains three corner nodes and one enriched meshfree node $\boldsymbol{x}_4$. Let be the affine transformation that maps the reference triangle onto to the triangle as depicted in Figure 1 and defined by

$$\boldsymbol{F}_e : \overline{\mathrm{T}}_e \rightarrow \mathrm{T}_e, \boldsymbol{x} = \boldsymbol{F}_e(\xi) = \left(F_e^1, F_e^2\right) = \left(\sum_{i=1}^{4} x_i \Psi_i(\xi,\eta),\ \sum_{i=1}^{4} y_i \Psi_i(\xi,\eta)\right)$$

$$\forall \boldsymbol{\xi} \in \overline{\mathrm{T}}_e \tag{1}$$

where $\boldsymbol{x} = [x, y]^T$, $\boldsymbol{\xi} = [\xi, \eta]^T$ and $\boldsymbol{P}_1\left(\overline{T}_e\right) = span\left\{\Psi_i, i = 1, \cdots, 4\right\}$ denotes the space contains a set of basis functions in $\overline{\mathrm{T}}_e$.

In Figure 1, the reference element is an equilateral triangle, with dark circles denoting the finite element node and open circles denoting the enriched meshfree node.

The location of the enriched meshfree node in reference element is given by

$$\left(\xi_4, \eta_4\right) = \left(\sum_{i=1}^{3} \xi_i / 3,\ \sum_{i=1}^{3} \eta_i / 3\right) \tag{2}$$

which is the centroid of the reference element. $\xi_i = \left(\xi_i, \eta_i\right), i = 1, 2, 3, 4$ are nodal co-coordinates of the reference element. The shape functions $\psi_i, i = 1, 2, 3, 4.$ of the reference element are constructed using a meshfree convex approximation.

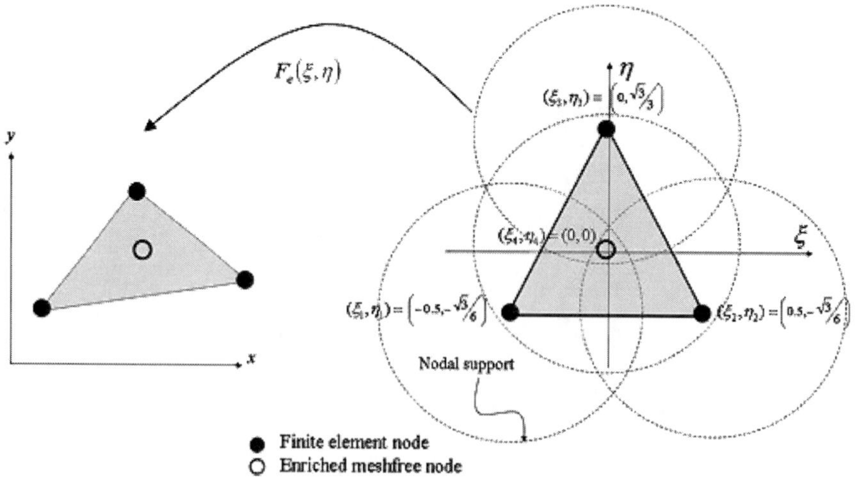

Figure 1. Isoparametric mapping in the 4-noded meshfree-enriched triangular finite element.

In this study, we employ the Generalized Meshfree Approximation (GMF) method [44] to obtain the meshfree convex approximation. The convex GMF approximation constructed using the inverse tangent basis function is denoted by GMF(*atan*). The cubic spline window function is chosen to be the weight function in GMF method.

In this study, each node in Figure 1 is assigned to a weight function with same circular support in the reference element $\overline{\mathrm{T}}_e$. The element mapping in the meshfree-enriched finite element method has been proven [45] to be bijective. In other words, the determinant of the Jacobian matrix computed using Equation (1) in the element mapping is positive everywhere in the element.

A detail derivation of GMF method and the corresponding mathematical properties can be found in [44].

Giving the four-noded ME-FEM shape functions, we define the following approximation space for the displacement field

$$\boldsymbol{V}^h(\Omega)=\left\{\boldsymbol{v}^h:\boldsymbol{v}^h\in\boldsymbol{H}_0^1(\Omega),\boldsymbol{v}^h\big|_{T_e}=\overline{\boldsymbol{v}}^h\circ\boldsymbol{F}_e^{-1},\overline{\boldsymbol{v}}^h\in\boldsymbol{P}_1\left(\overline{T}_e\right)\ \forall\,\mathrm{T}_e\in M_h\right\} \tag{3}$$

which consists of functions in Sobolev space $\boldsymbol{H}^1(\Omega)$ vanish on the boundary. Since the shape functions constructed using the GMF method are convex, they exhibit the following convexity properties which are not mutually $(\cdot,\cdot)_{V^h}$-orthonormal

$$\begin{aligned}
&\Psi_i\left(\xi_j\right)=\delta_{ij}\quad 1\le i,j\le 3\\
&\Psi_4\left(\xi\right)>0\ \ \forall\xi\in\overline{\mathrm{T}}_e\\
&\Psi_4\left(\xi\right)=0\ \ \forall\xi\in\partial\overline{\mathrm{T}}_e\\
&\Psi_i\left(\xi_4\right)>0\ \ 1\le i\le 3\ \ \textit{and}\ \ \Psi_4\left(\xi_4\right)\ne 1
\end{aligned} \tag{4}$$

Figure 2 shows the shape functions of a four-noded ME-FEM element using GMF(*atan*) approximation where the Kronecker-delta property is satisfied at the boundary [45]. The shape functions of the ME-FEM element reduce to the standard linear finite element shape functions along the element edge that exhibit the convex approximation property.

2.2. Smoothing of Deformation Gradient in Meshfree-Enriched Finite Element Triangular Elements

In order to provide a locking-free analysis for the elastomers using meshfree-enriched triangular elements, an area-weighted strain smoothing scheme originally introduced by Wu and Hu [45] in the near-incompressible linear elastic problem is adopted in this study for the smoothing of deformation gradient for the nonlinear hyperelastic problem. The strain smoothing technique was firstly proposed by Chen and his research group [15] and has been widely used in meshfree methods to enforce the linear exactness in Galerkin approach as well as to improve the computational efficiency.

a b

[a]shape function of corner node.
[b]shape function of central node.

Figure 2. GMF(atan) convex approximation in a 4-noded meshfree-enriched finite element.

The strain smoothing technique in meshfree methods was also applied to finite element method providing a softening effect to improve the solution accuracy and led to various node-based, element-based and edge-based smoothed finite element formulations of Liu and his co-authors [10, 23, 28]. The deformation gradient smoothing scheme is described in the following:

Let be the three vertices of a triangular element $T_j \in M_h$. is the centroid of the triangular element T_j. We connect to the three vertices of the triangle by straight lines to divide the triangle into three sub-triangles S_l, S_m *and* S_n as shown in Figure 3. Each sub-triangle shares the element edge *m* of the triangle and carries one integration point. Since each sub-triangle occupies the same area, all three integration points are assigned to the same weight for the numerical integration. The location of the integration points $\xi_{gi}, 1 \le i \le 3$ is chosen to form a symmetric pattern in the reference co-coordinate system such that for the enriched meshfree node *I*, one has

$$\sum_{i=1}^{3} \frac{\partial \Psi_I\left(\xi_{gi}\right)}{\partial \xi} = 0 \tag{5}$$

and

$$\sum_{i=1}^{3} \frac{\partial \Psi_I\left(\xi_{gi}\right)}{\partial \eta} = 0 \tag{6}$$

By doing that, the integration point of each sub-triangle can be chosen to any point locating on the extension of the straight line associated with that sub-triangle as illustrated by the dashed line in Figure 3. For simplicity, we choose the integration point to locate at the end of the extension line which is the midpoint of the element edge in this study. An eigenvalue analysis verifies that the proposed ME-FEM triangular element with this three-point integration rule does not contain spurious zero energy modes.

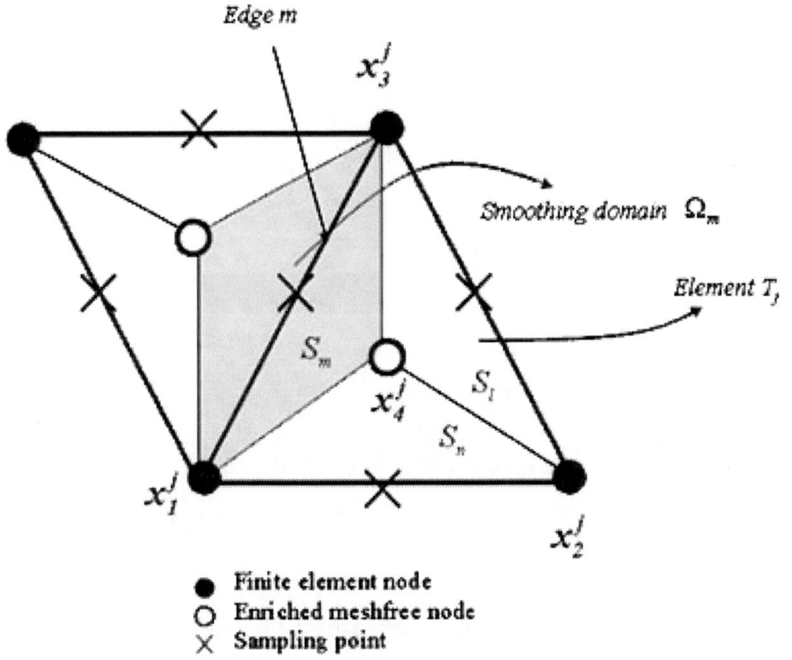

Figure 3. Strain smoothing in ME-FEM triangle elements.

The smoothing domain corresponding to the edge m for the adjacent elements is defined as and the smoothed deformation gradient is given in terms of its components by

$$\bar{F}_{ij}\left(\boldsymbol{X}_{gk}\right)=\frac{1}{A_m}\int_{\Omega_m} F_{ij}\left(\boldsymbol{X}_{gk}\right)\Phi_m\left(\boldsymbol{X}\right)d\Omega \tag{7}$$

where is the area of the smoothing domain Ω_m, X_{gk} ,k=1,2,3 is the integration point of the element that corresponds to the ξ_{gk} of $\overline{\mathrm{T}}_j$. u_i^h is the approximation of displacement in i-component and is obtained by the linear combination of Lagrangian shape function of meshfree-enriched triangular element. It can be expressed by

$$u_i^h\left(\boldsymbol{X}\right)=\sum_{I=1}^{4}\Psi_I\left(\boldsymbol{X}\right)u_{iI}\quad, i=1,2 \tag{8}$$

where u_{iI} is the nodal coefficient needs to be determined. in Equation (7) is the characteristic or smoothing function of the smoothing domain defined by

$$\Phi_m\left(\boldsymbol{x}\right)=\begin{cases}1, & if\ \boldsymbol{x}\in\Omega_m\\ 0, & else\end{cases} \tag{9}$$

Subsequently, Equation (7) can be rewritten using Equation (8) to yield

$$\bar{F}_{ij}^h = \frac{1}{A_m}\int_{\Omega_m}\left(\frac{\partial u_i^h}{\partial X_j}+\delta_{ij}\right)\Phi_m(X)\,d\Omega = \frac{1}{A_m}\int_{\Omega_m}\left(\frac{\partial u_i^h}{\partial X_j}\right)\Phi_m(X)\,d\Omega+\delta_{ij}$$
$$= \frac{1}{A_m}\int_{\Omega_m}\sum_{I=1}^{4}\left(\frac{\partial \Psi_I}{\partial X_j}u_{iI}\right)\Phi_m(X)\,d\Omega+\delta_{ij} \equiv \bar{g}_{ij}^h+\delta_{ij} \tag{10}$$

where

$$\bar{g}_{ij}^h = \frac{1}{A_m}\int_{\Omega_m}\sum_{I=1}^{4}\left(\frac{\partial \Psi_I}{\partial X_j}u_{iI}\right)\Phi_m(X)\,d\Omega \tag{11}$$

Since the smoothed deformation gradient is defined locally on each smoothing domain and no continuity conditions are applied to the boundaries of Ω_m, the approximation space of smoothed deformation gradient can be defined by

$$\boldsymbol{\Xi}^h(\Omega)=\left\{\bar{\boldsymbol{F}}^h : \bar{\boldsymbol{F}}^h \in L^2(\Omega), \bar{\boldsymbol{F}}^h \text{ contains piecewise constants } \forall\, \Omega_m \in M_h\right\} \tag{12}$$

In linear elastic problem, an assumed strain method [37] can be employed to formulate the discrete equations as given in [45] where pressure is calculated via post-processing from the following constitutive relation

$$p^h = -\lambda\, \overline{div\left(\boldsymbol{u}^h\right)} = -\lambda tr\, \bar{\varepsilon}\left(\boldsymbol{u}^h\right) = -\lambda\frac{1}{A_m}\int_{\Omega_m} tr\left[\varepsilon\left(\boldsymbol{u}^h\right)\right]\Phi_m(\boldsymbol{X})\,d\Omega \tag{13}$$

The symbol is the Lamé constant or bulk modulus which is related to the Young's modulus E and Poisson ratio v by

$$\lambda = (vE)/\left[(1+v)(1-2v)\right].$$

Although the pressure does not directly involve in the computation, the well-posedness of the problem in near-incompressible regime is still subject to a stability condition between the displacement space and an implicit pressure space induced by Equation (13).

Alternatively, an equivalent mixed formulation was derived by Wu and Hu [45] for the stability study of meshfree-enriched finite element method in the near-incompressible limit. Their numerical inf-sup results indicate the pair of spaces ($\boldsymbol{V}^h$, $\boldsymbol{P}^h$) is inf-sup stable.

3. NONLINEAR MESHFREE-ENRICHED FINITE ELEMENT FORMULATION

3.1. Elasticity

3.1.1. Modified Variational Functional and Material Nonlinearity

To introduce the smoothed deformation gradient into Galerkin approximation in the large deformation analysis, the following mixed three-field Hu-Washizu-de Veubeke energy functional [42] is considered for the derivation of nonlinear finite element formulation for hyperelastic materials.

$$U_{HW}\left(\boldsymbol{u},\bar{\boldsymbol{F}},\bar{\boldsymbol{\tau}}\right)=\int_{\Omega}W\left(\bar{\boldsymbol{F}}\right)d\Omega+\int_{\Omega}\bar{\boldsymbol{\tau}}:\left(\nabla_0\boldsymbol{u}-\bar{\boldsymbol{F}}\right)d\Omega-W_{ext}\left(\boldsymbol{u}\right) \qquad (14)$$

where the displacements $\boldsymbol{u}$, smoothed deformation gradient and smoothed first Piola-Kirchhoff stresses $\bar{\boldsymbol{\tau}}$ are independently varied. The symbol denotes the gradient operator with respect to the original configuration. is the assumed strain energy density function. The term designates the external work. The smoothed first Piola-Kirchhoff stresses is related to the smoothed deformation gradient by

$$\bar{\boldsymbol{\tau}}=\frac{\partial W}{\partial\bar{\boldsymbol{F}}} \qquad (15)$$

A strain energy density function W with decomposition of deviatoric and volumetric parts is adopted [11, 12] for this study and is given in the following

$$W\left(\bar{I}_1,\bar{I}_2,J\right)=\hat{W}\left(\hat{I}_1,\hat{I}_2\right)+\tilde{W}\left(\bar{J}\right) \qquad (16)$$

where the smoothed deviatoric energy density function and the smoothed volumetric energy density function are defined by

$$\hat{W}\left(\hat{I}_1,\hat{I}_2\right)=\sum_{p+q}^{\infty}A_{pq}\left(\hat{I}_1-3\right)^p\left(\hat{I}_2-3\right)^q \qquad (17)$$

$$\tilde{W}\left(\bar{J}\right)=\frac{\lambda}{2}\left(\bar{J}-1\right)^2 \qquad (18)$$

$\hat{I}_1$ and are the reduced invariants [34] that represent pure deviatoric deformation respectively and are defined by

$$\hat{I}_1=I_1I_3^{-1/3} \qquad (19)$$

$$\hat{I}_2 = I_2 I_3^{-2/3} \tag{20}$$

where I_1, I_2 and I_3 are the first, second and third invariants of the smoothed Green deformation tensor $\overline{G}_{ij} = \overline{F}_{ki}\overline{F}_{kj}$. $\overline{J} = \det(\overline{\boldsymbol{F}}) = I_3^{1/2}$ is a measure of volumetric deformation.

Equation (18) can be viewed as a penalty formulation of incompressible hyperelasticity. In Mooney-Rivlin model [11], Equation (17) is given by

$$\hat{W}(\hat{I}_1, \hat{I}_2) = A_{10}(\hat{I}_1 - 3) + A_{01}(\hat{I}_2 - 3) \tag{21}$$

Subsequently, the smoothed first Piola-Kirchhoff stresses $\overline{\boldsymbol{\tau}}$ in Equation (15) can be obtained by

$$\overline{\tau}_{ij} = 2\overline{F}_{ki}\left[\hat{K}_1 I_3^{-1/3}\left(\delta_{kj} - \frac{1}{3}I_1\overline{G}_{kj}^{-1}\right) + \hat{K}_2 I_3^{-2/3}\left(I_1\delta_{kj} - \overline{G}_{kj} - \frac{2}{3}I_2 G_{kj}^{-1}\right)\right] + pJ\overline{F}_{ki}\overline{G}_{kj}^{-1} \tag{22}$$

where

$$\hat{K}_i = \frac{\partial \hat{W}}{\partial \overline{I}_i}, i = 1,2 \tag{23}$$

The pressure p in Equation (22) is related to the smoothed volumetric energy density function through Equation (18) and is given by

$$p = tr(\overline{\boldsymbol{\tau}}) = \frac{\partial \widetilde{W}(\overline{J})}{\partial \overline{J}} = \lambda(\overline{J} - 1) \tag{24}$$

Equation (24) implies an imposition of incompressible constraint through the penalty method whereas the bulk modulus is considered as a penalty number used to mimic the incompressibility of elastomers.

Note that no pressure filtering or smoothing are required for the proposed method to extract any possible checkerboard modes from the implicit pressure space induced from Equation (24). The behavior of hydrostatic pressure computed from Equation (24) is further studied in the numerical examples.

The discretized total energy functional corresponds to Equation (14) can be written as:

$$U_{HW}^h : \boldsymbol{V}^h \times \boldsymbol{\Xi}^h \times \boldsymbol{\Theta}^h \rightarrow R$$
$$U_{HW}^h(\boldsymbol{u}^h, \overline{\boldsymbol{F}}^h, \overline{\boldsymbol{\tau}}^h) = \int_\Omega W(\overline{\boldsymbol{F}}^h)d\Omega + \int_\Omega \overline{\boldsymbol{\tau}}^h : (\nabla_0 \boldsymbol{u}^h - \overline{\boldsymbol{F}}^h)d\Omega - W_{ext}(\boldsymbol{u}^h) \tag{25}$$

Since the space $\boldsymbol{\Xi}^h$ of smoothed deformation gradient and the space $\boldsymbol{\Theta}^h \in L^2$ of smoothed first Piola-Kirchhoff stresses contain piecewise constants in Ω, the second term on the RHS of Equation (25) can be further expressed using Equation (7) to yield

$$\int_\Omega \bar{\boldsymbol{\tau}}^{hT}\left(\nabla_0 \boldsymbol{u}^h - \bar{\boldsymbol{F}}^h\right)d\Omega = \sum_{m=1}^{nm} \bar{\boldsymbol{\tau}}^{hT} \int_{\Omega_M}\left(\nabla_0 \boldsymbol{u}^h - \bar{\boldsymbol{F}}^h\right)d\Omega = \sum_{m=1}^{nm} \bar{\boldsymbol{\tau}}^{hT}\left(\int_{\Omega_M} \nabla_0 \boldsymbol{u}^h d\Omega - \bar{\boldsymbol{F}}^h A_m\right)$$
$$= \sum_{m=1}^{nm} \bar{\boldsymbol{\tau}}^{hT}\left(\int_{\Omega_M} \nabla_0 \boldsymbol{u}^h d\Omega - \int_{\Omega_M} \nabla_0 \boldsymbol{u}^h d\Omega\right) = 0 \tag{26}$$

The index nm in Equation (26) denotes the total number of smoothing domain Ω_m in Ω ($\Omega = \cup_m \Omega_m$). Equation (26) implies an orthogonal condition between the stress field $\bar{\boldsymbol{\tau}}^h$ and the difference of the deformation gradient $\nabla_0 \boldsymbol{u}^h$ field and the smoothed strain field $\bar{\boldsymbol{F}}^h$. This is equivalent to the assumed strain variation principle of Simo and Hughes [37].

After eliminating the stress components from Equation (25) using Equation (26) the following modified Hu-Washizu functional is obtained depending only on displacement and smoothed deformation fields:

$$U^h_{HW\,\mathrm{mod}}\left(\boldsymbol{u}^h, \bar{\boldsymbol{F}}^h\right) = \int_\Omega W\left(\bar{\boldsymbol{F}}^h\right)d\Omega - W_{ext}\left(\boldsymbol{u}^h\right) \tag{27}$$

3.1.2. Nonlinear Finite Element Formulation and Discrete Equation

Since the material frame indifference restricts the dependence of strain energy density function W on smoothed deformation gradient $\bar{\boldsymbol{F}}$, it is advantageous to express the variational equation in terms of the total Lagrangian formulation. Linearization of potential energy in Equation (27) leads to the following variational equation:

$$\delta U^h_{HW\,\mathrm{mod}}\left(\boldsymbol{u}^h, \bar{\boldsymbol{F}}^h\right) = \int_\Omega \delta \bar{F}^h_{ij} \frac{\partial W\left(\bar{\boldsymbol{F}}^h\right)}{\partial \bar{F}_{ij}} d\Omega - \delta W_{ext}\left(\boldsymbol{u}^h\right) = \int_\Omega \delta \bar{g}^h_{ij} \bar{\tau}_{ij}\left(\bar{\boldsymbol{F}}^h\right)d\Omega - \delta W_{ext}\left(\boldsymbol{u}^h\right) \tag{28}$$

where is defined in Equation (11).

Linearization of Equation (28) yields

$$\Delta\delta U^h_{HW\,\mathrm{mod}}\left(\boldsymbol{u}^h, \bar{\boldsymbol{F}}^h\right) = \int_\Omega \delta \bar{g}^h_{ij}\left[\frac{\partial^2 W\left(\bar{\boldsymbol{F}}^h\right)}{\partial \bar{F}_{ij}\partial \bar{F}_{kl}} + \delta_{ik} \bar{F}^h_{mj} \bar{\tau}_{ml}\left(\bar{\boldsymbol{F}}^h\right)\right]\Delta g^h_{kl} d\Omega - \Delta\delta W_{ext}\left(\boldsymbol{u}^h\right)$$
$$= \int_\Omega \delta \bar{g}^h_{ij}\left[\bar{C}_{ijkl}\left(\bar{\boldsymbol{F}}^h\right) + \bar{S}_{ijkl}\left(\bar{\boldsymbol{F}}^h\right)\right]\Delta g^h_{kl} d\Omega - \Delta\delta W_{ext}\left(\boldsymbol{u}^h\right) \tag{29}$$

where

$$\bar{C}_{ijkl}\left(\bar{\boldsymbol{F}}^h\right) = \frac{\partial^2 W\left(\bar{\boldsymbol{F}}^h\right)}{\partial \bar{F}_{ij}\partial \bar{F}_{kl}}, \tag{30}$$

$$\bar{S}_{ijkl}\left(\bar{\boldsymbol{F}}^h\right)=\delta_{ik}\bar{F}^h_{mj}\bar{\tau}_{ml}\left(\bar{\boldsymbol{F}}^h\right) \tag{31}$$

Now introducing the approximations of displacement in Equation (8) and smoothed deformation gradient in Equation (10) into Equation (29) together with the numerical integration described in Section 2.2 to yield the following incremental matrix equation

$$\delta\{\boldsymbol{U}\}^T\bar{\boldsymbol{K}}^v_{n+1}(\Delta\{\boldsymbol{U}\})^{v+1}_{n+1}=-\delta\{\boldsymbol{U}\}^T\boldsymbol{R}^v_{n+1} \tag{32}$$

where is the displacement vector. The tangent stiffness matrix $\boldsymbol{K}^v_{n+1}(\Delta\{\boldsymbol{U}\})^{v+1}_{n+1}$ contains the material and geometric stiffness matrices that evaluated at the v-th iteration during the $(n+1)$ time incremental step. is the residual nodal force vector. They are given by

$$\bar{\boldsymbol{K}}_{IJ}=\int_{\Omega}\bar{\boldsymbol{B}}_I^T\left[\bar{\boldsymbol{C}}\left(\bar{\boldsymbol{F}}^h\right)+\bar{\boldsymbol{S}}\left(\bar{\boldsymbol{F}}^h\right)\right]\bar{\boldsymbol{B}}_J d\Omega \tag{33}$$

$$-\boldsymbol{R}=\boldsymbol{f}^{ext}-\boldsymbol{f}^{int} \tag{34}$$

$$\boldsymbol{f}_I^{int}=\int_{\Omega}\bar{\boldsymbol{B}}_I^T\bar{\boldsymbol{\tau}}\left(\bar{\boldsymbol{F}}^h\right)d\Omega \tag{35}$$

$$\boldsymbol{f}_I^{ext}=\int_{\Omega}\Psi_I\boldsymbol{b}_0 d\Omega+\int_{\Gamma^h}\Psi_I\boldsymbol{h}_0 d\Gamma \tag{36}$$

where is the body force density measured in the original configuration and is the prescribed traction on the original boundary Γ^h. is the smoothed gradient matrix which is under plain strain assumption and given by

$$\bar{\boldsymbol{B}}_I=\frac{1}{A_m}\begin{bmatrix}\int_{\Omega_m}\left(\frac{\partial\Psi_I}{\partial X_1}\right)\Phi_m(X)d\Omega & 0\\ 0 & \int_{\Omega_m}\left(\frac{\partial\Psi_I}{\partial X_2}\right)\Phi_m(X)d\Omega\\ \int_{\Omega_m}\left(\frac{\partial\Psi_I}{\partial X_2}\right)\Phi_m(X)d\Omega & 0\\ 0 & \int_{\Omega_m}\left(\frac{\partial\Psi_I}{\partial X_1}\right)\Phi_m(X)d\Omega\end{bmatrix} \tag{37}$$

Because the smoothed gradient matrix is defined in the original configuration, the smoothing procedure described in Section 2.2 only needs to be performed at once. The smoothed first Piola-Kirchhoff stresses is given in Equation (22) and is expressed in vector form with its components given by

$$\bar{\boldsymbol{\tau}}=\{\bar{\tau}_{11},\bar{\tau}_{22},\bar{\tau}_{12},\bar{\tau}_{21}\}^T \tag{38}$$

The discrete hydrostatic pressure can be obtained according to Equation (24)

$$p^h = -\lambda\left(\bar{J}^h - 1\right) = -\lambda\left(\frac{1}{A_m}\int_{\Omega_m} J^h \Phi_m(\mathbf{X})d\Omega\right) \tag{39}$$

where the determinant of the smoothed deformation gradients is evaluated through Equation (10) Although the pressure does not directly involve in the nonlinear computation, the stability of the nonlinear formulation is still subject to the inf-sup pair requirements for displacement field and the implicit pressure field as in the linear analysis.

Since the interpolation of meshfree-enriched triangular element preserves Kronecker-delta property at the element boundary [45], the essential boundary conditions can be treated in the standard way.

From Equation (4) it is not difficult to see that the computed displacements of meshfree-enriched nodes are not the real nodal displacements. This is because the shape function of the enriched node does not possess Kronecker-delta property. According to Equation (8), the real nodal displacement $u_i^h\left(\mathbf{X}_J\right), i = 1, 2$ of the meshfree-enriched node *J* can be obtained by

$$u_i^h\left(\mathbf{X}_J\right) = \hat{u}_{iJ} = \sum_{I=1}^{4} \Psi_I\left(\mathbf{X}_J\right) u_{iI}, i = 1, 2 \tag{40}$$

where

$$\hat{u}_{iJ} \neq u_{iJ}, i = 1, 2 \tag{41}$$

3.2. Plasticity

For path-independent material, the energy function (14) is established in the current configuration using the updated Lagrangian formulation. By introducing the Lagrangian strain smoothing, we can obtain the modified variational equation similar as in the assumed strain method:

$$\delta U^h = \int_\Omega \delta \bar{u}_{i,j}^h \bar{\sigma}_{ij} d\Omega - \delta W_{ext} \tag{42}$$

where is the smoothed Cauchy stress, is the smoothed strain, is the spatial coordinate calculated from material coordinate and displacement $\boldsymbol{u}\left(\boldsymbol{X}\right)$, and represents the domain of the current configuration. The linearization of Equation (42) is

$$\Delta\delta U^h = \int_\Omega \delta \bar{u}_{i,j}^h \left(\bar{C}_{ijkl}^\sigma + \bar{T}_{ijkl}^\sigma\right) \Delta \bar{u}_{k,l}^h d\Omega - \Delta\delta W_{ext} \tag{43}$$

where is the material response tensor and is geometric response tensor. Considering that the ME-FEM shape function and the smoothed gradient of displacement approximation are defined in the original configuration, we transform the variational equations (42) and (43) to the undeformed configuration as

$$\delta U^h = \int_{\Omega_0} \overline{\frac{\partial \delta u_i^h}{\partial X_k}} \bar{F}_{kj}^{-1} \bar{\sigma}_{ij} J_0 d\Omega_0 - \delta W_{ext} \tag{44}$$

$$\Delta\delta U^h = \int_{\Omega_0} \overline{\frac{\partial \delta u_i^h}{\partial X_m}} \bar{F}_{mj}^{-1} \left(\bar{C}_{ijkl}^{\sigma} + \bar{T}_{ijkl}^{\sigma} \right) \bar{F}_{nl}^{-1} \overline{\frac{\partial \Delta u_k^h}{\partial X_n}} J_0 d\Omega_0 - \Delta\delta W_{ext} \tag{45}$$

where is undeformed domain, $J_0 = \det(\boldsymbol{F})$. is the smoothed deformation gradient. In the computation, in Eqs. (44) and (45) is calculated only at the integration points, which can be obtained by taking the direct inverse of pointwise.

The smoothed incremental strain at the integration point is computed by

$$\Delta\bar{u}_{i,j}\left(\boldsymbol{X}_{gk}\right) = \overline{\frac{\partial \Delta u_i}{\partial X_m}} \bar{F}_{mj}^{-1} = \Delta\bar{F}_{im}\left(\boldsymbol{X}_{gk}\right) \bar{F}_{mj}^{-1}\left(\boldsymbol{X}_{gk}\right) \tag{46}$$

where is smoothed deformation gradient computed by Equation (10). Subscribing Eqs. (10) and (46) into Equation (45) leads to the following linearization of variational equation:

$$\Delta\delta U^h = \int_{\Omega_0} \delta\bar{g}_{im}^h \bar{F}_{mj}^{-1} \left[\bar{C}_{ijkl}^{\sigma}\left(\bar{\boldsymbol{F}}^h\right) + \bar{T}_{ijkl}^{\sigma}\left(\bar{\boldsymbol{F}}^h\right) \right] \bar{F}_{nl}^{-1} \Delta\bar{g}_{kn}^h \bar{J}_0\left(\bar{\boldsymbol{F}}^h\right) d\Omega_0 - \Delta\delta W_{ext} \tag{47}$$

Introducing the approximations of displacement in Equation (8) and smoothed deformation gradient in Equation (10) into Equation (47) yields the stiffness matrix and internal force vector:

$$\bar{\boldsymbol{K}}_{IJ} = \int_{\Omega_0} \bar{\boldsymbol{B}}_I^T \bar{\boldsymbol{G}}^T \left[\bar{\boldsymbol{C}}^{\sigma}\left(\bar{\boldsymbol{F}}^h\right) + \bar{\boldsymbol{T}}^{\sigma}\left(\bar{\boldsymbol{F}}^h\right) \right] \bar{\boldsymbol{G}} \bar{\boldsymbol{B}}_J \bar{J}_0 d\Omega_0 \tag{48}$$

$$\boldsymbol{f}_I^{int} = \int_{\Omega_0} \bar{\boldsymbol{B}}_I^T \bar{\boldsymbol{G}}^T \bar{\boldsymbol{\sigma}}\left(\bar{\boldsymbol{F}}^h\right) \bar{J}_0 d\Omega_0 \tag{49}$$

where

$$\bar{\boldsymbol{G}} = \begin{bmatrix} \bar{\boldsymbol{F}}_{11}^{-1} & 0 & \bar{\boldsymbol{F}}_{21}^{-1} & 0 \\ 0 & \bar{\boldsymbol{F}}_{22}^{-1} & 0 & \bar{\boldsymbol{F}}_{12}^{-1} \\ \bar{\boldsymbol{F}}_{12}^{-1} & \bar{\boldsymbol{F}}_{21}^{-1} & \bar{\boldsymbol{F}}_{22}^{-1} & \bar{\boldsymbol{F}}_{11}^{-1} \end{bmatrix} \tag{50}$$

$$\bar{\boldsymbol{\sigma}} = \begin{bmatrix} \bar{\sigma}_{11} \\ \bar{\sigma}_{22} \\ \bar{\sigma}_{12} \end{bmatrix} \tag{51}$$

and is the smoothed gradient matrix computed by Equation (37).

4. Numerical Examples

In this section, the nonlinear performance of the meshfree-enriched triangular elements with area-weighted strain smoothing (ME-Tri-AW) is evaluated through the following numerical examples in the two-dimensional case. As comparison, we also provide the results using (1) the standard generalized meshfree method with Gauss integration (GMF-GI) [44] (2) the 3-noded triangular element with nodal integration (NICE-T3) [26].

As in the ME-Tri-AW, the meshfree shape functions in GMF-GI are constructed using the Generalized Meshfree Approximation (GMF) method [44]. The weight function in GMF method is chosen to be cubic B-spline kernel function for both ME-Tri-AW and GMF-GI. In order to relieve the possible volumetric locking in GMF-GI, a large normalized support size of 1.95 is adopted in the kernel function.

A number larger than 1.95 for the normalized support size would cause numerical instability in GMF-GI, therefore it is not considered in this study. On the other hand, the result in ME-Tri-AW is insensitive to the normalized support size (ranged from 1.01 to 10.01 or even larger).

Therefore we simply select the normalized support size to be 1.1 for ME-Tri-Aw in all three examples.

The computational domain of each example is discretized by a set of triangular elements whereas additional meshfree nodes are added into each triangular element in ME-Tri-AW as a way described in Section 2. In GMF-GI, each triangular element also serves as the background cell for the numerical integration using one-point integration rule.

Since the Kronecker-delta property of meshfree shape functions is preserved in the Generalized Meshfree Approximation method, no special essential boundary condition treatments are required for ME-Tri-AW and GMF-GI.

A standard Newton-Raphson method is employed to solve the nonlinear equation (32). If the number of iterations reaches the maximum number of allowed equilibrium iterations, the current time step is reset and the iteration with a reduction of time increment by a scale factor of 0.9 is redoing. This execution repeats until the convergence is achieved. If the number of resets in single time step reaches its maximum of 10, the execution is terminated.

4.1. Radial Bushing of Rubber Cylinder

This example is studied to identify the applicability of the proposed method to highly constrained problem. The radial bushing problem, as shown in Figure 4 (a), is analyzed, where the outer surface is fixed, and the inner surface of rubber cylinder sticks with an un-deformed core moving along vertical direction.

The hyperelastic characteristic of elastomers is described by the Mooney-Rivlin strain energy density function in Equation (21) with material constants [12] given by

$$\begin{aligned} A_{10} &= 0.2599\text{MPa} \\ A_{01} &= 0.1608\text{MPa} \end{aligned} \tag{52}$$

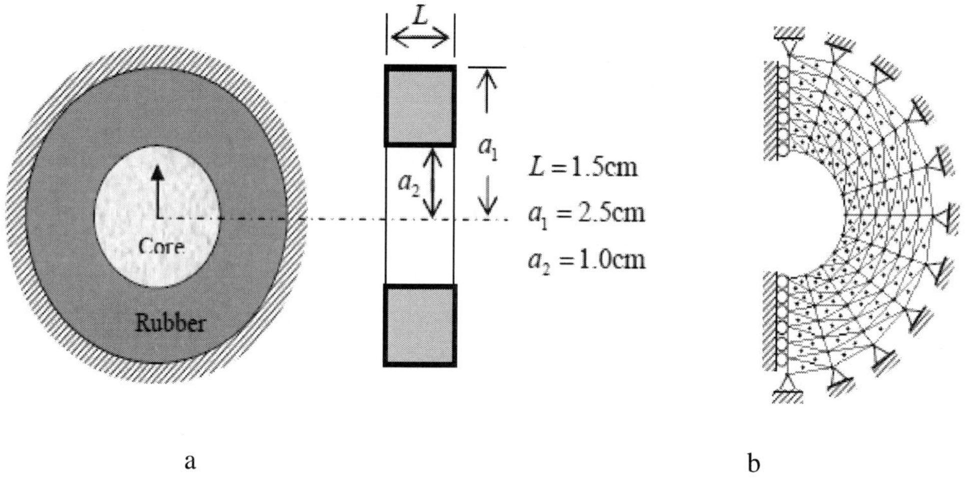

Figure 4. Rubber bushing: (a) problem description; (b) half model discretized by meshfree enriched triangular elements (ME-Tri-AW).

The bulk modulus is taken to be which is corresponding to the Poisson ratio in the linear elasticity problem. Under the plane strain assumption, the linear relationship between radial force and displacement has been studied by Stevenson [40] as follows

$$F = \frac{8\pi\left(A_{10} + A_{01}\right)\left(\beta^2 + 1\right)Ld}{\left(\beta^2 + 1\right)\ln\beta - \left(\beta^2 - 1\right)} \tag{53}$$

where $\beta = a_1/a_2$, and are the outer and inner radius of the un-deformed cylinder, and L is the shape factor, as shown in Figure 4 (a). Considering the symmetry of the problem, we model half of the 2D cross-section by 120 triangular elements. The discretization of ME-Tri-AW with enriched nodes is plotted in Figure 4 (b).

The load-displacement curves generated by three numerical methods are shown in Figure 5. ME-Tri-AW result shows a linear load-displacement growth within the bushing operating range, which agrees well with analytical solution.

Nonlinear response in ME-Tri-AW result occurs as the top inner surface is pushed very close to the outer surface. GMF-GI result exhibits severe volumetric locking and fails to deliver desirable force prediction to the rubber bushing. On the other hand, the nonlinear iteration of NICE-T3 is not able to advance when the radial displacement is beyond 0.2cm.

Superior performance of ME-Tri-AW over the other two methods is demonstrated in the displacement field in this highly constrained problem. Figure 6 (a) shows the plots of maximum attainable deformation in three methods. It is observed that ME-Tri-AW provides the smoothest displacement field among three methods, while spurious modes emerge in NICE-T3 result.

These spurious deformation modes appear even in the early stage and eventually cause convergence difficulties in NICE-T3. The deformation history using ME-Tri-AW is provides in Figure 6 (b).

The hydrostatic pressure contour plots of three methods are shown in Figure 7. As in the previous example, the GMF-GI exhibits a strong pressure oscillation.

A high level of hydrostatic pressure field in GMF-GI result indicates a severe locking behavior and the predicted response is overly stiff. Although the oscillation does not show up in the NICE-T3 result, the pressure field of the same sign implies a divergence in displacement field caused by the spurious deformation modes.

Over all, ME-Tri-AW provides the best solution in three methods as it is simultaneously free of locking and pressure oscillation.

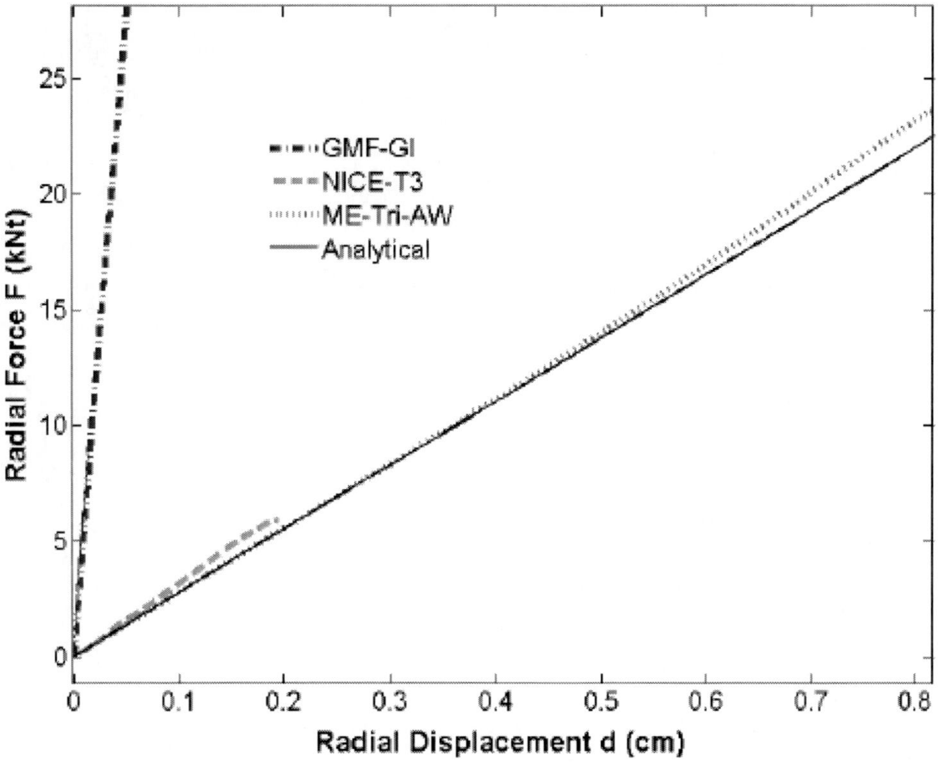

Figure 5. Load-displacement curves of rubber bushing problem.

GMF-GI ($d = 0.3$cm) NICE-T3($d = 0.15$cm) ME-Tri-AW($d = 0.75$cm)

a

ME-Tri-AW ($d = 0.3$cm) ME-Tri-AW ($d = 0.45$cm) ME-Tri-AW($d = 0.60$cm)

b

Figure 6. Deformation of rubber insertion.

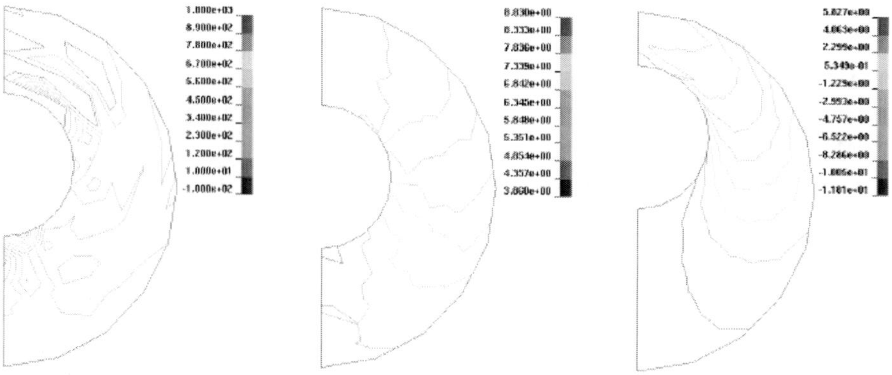

GMF-GI ($d = 0.3$cm) NICE-T3($d = 0.15$cm) ME-Tri-AW($d = 0.75$cm)

Figure 7. Hydrostatic pressure contour for rubber bushing problem.

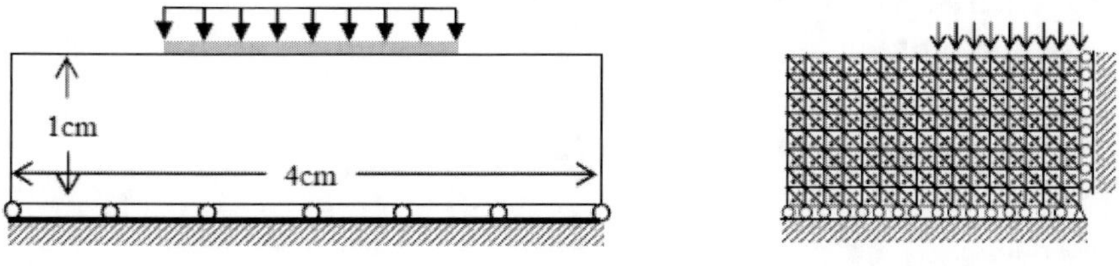

Figure 8. Compression problem: (a) compression model; (b) half model discretized by meshfree-enriched triangular elements.

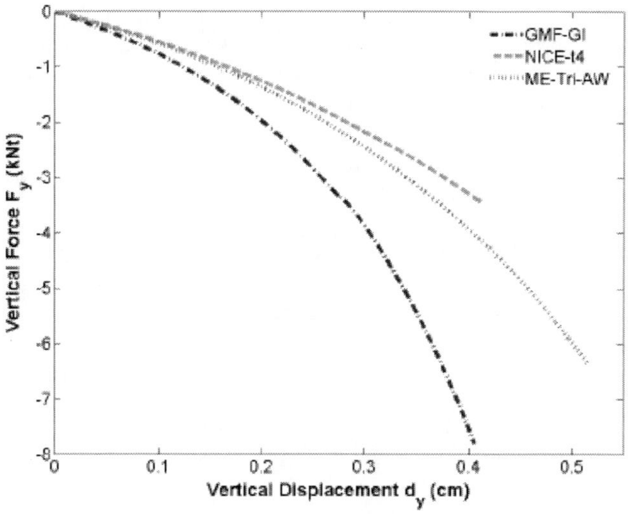

Figure 9. Load-displacement curves of compression problem.

4.2. Compression Problem

In this example, a compression test of elastomer is conducted to examine the applicability of the proposed method to the rubber seal analysis. The problem statement and boundary conditions of the compression problem are given in Figure 8 (a).

Based on the symmetry of this problem, half model is used in the analysis. The model is discretized by 256 triangular elements. The meshfree-enriched finite element discretization is shown in Figure 8(b). The same hyperelastic material is used as in example 4.1.

The load-displacement curves are shown in Figure 9. NICE-T3 and ME-Tri-AW predict similar load-displacement curves on the initial slope. NICE-T3 behaves softly in the nonlinear range and experiences lower rates of convergence due to excessive spurious deformation modes. On the other hand, GMF-GI responses overly stiff when it is compared with NICE-T3 and ME-Tri-AW. A large support size may be used to improve the volumetric locking problem in GMF-GI. However this would cause the ill-conditioning problem. As a result, divergence happens even earlier. A convergence study using ME-Tri-AW is conducted and the load-displacement results, as shown in Figure 10, indicate the convergence of ME-Tri-AW as the mesh is refined.

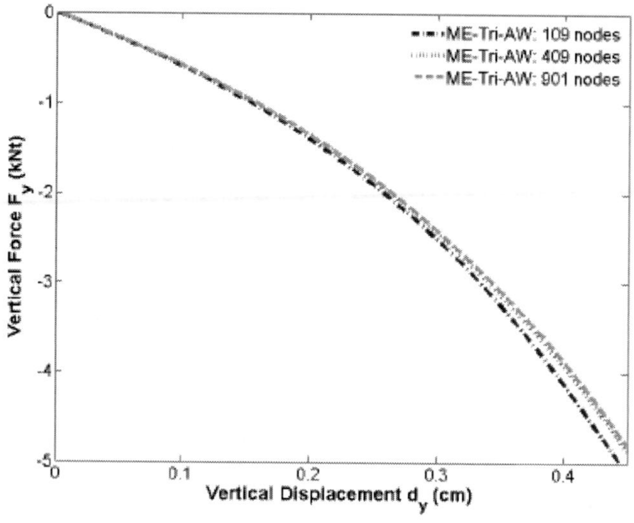

Figure 10. Convergence study of compression problem using ME-Tri-AW.

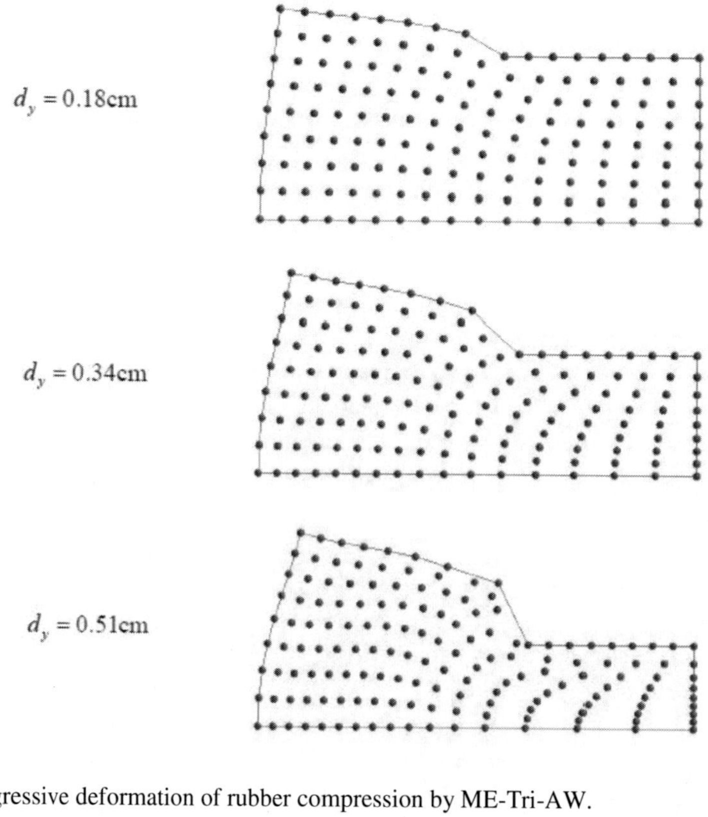

Figure 11. Progressive deformation of rubber compression by ME-Tri-AW.

Figure 11 shows the deformation history of ME-Tri-AW result plotted on the finite element nodes in three different loading steps.

The volumetric locking solution of GMF-GI reflects on the oscillation problem in its pressure field as shown in Figure 12. The pressure field of NICE-T3, on the other hand, is not reliable due to the existence of spurious deformation modes. Both GMF-GI and NICE-T3 have almost the same compressibility. In contrast, ME-Tri-AW presents the best compressibility among three methods. As in the previous two examples, this test suggests that the proposed method is free of locking and oscillation despite the use of a small support size in the finite deformation range.

4.3. Punch Problem

The punch example considers an elaso-plastic material subjected to displacement-controlled rigid footing under highly constrained boundaries as shown in Figure 13.

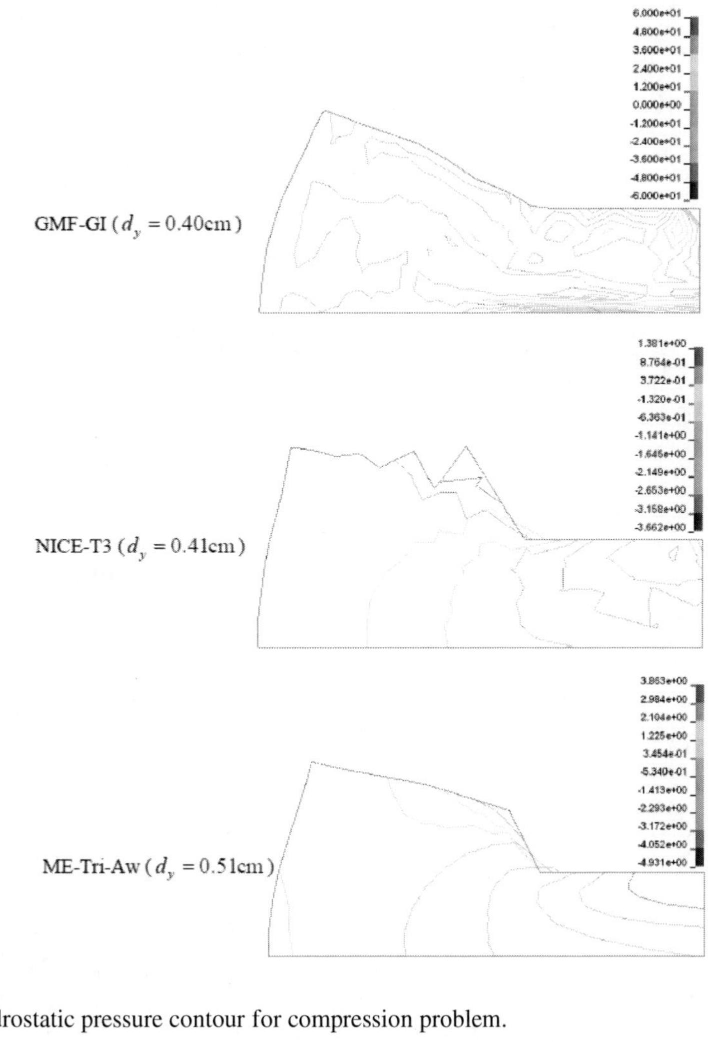

Figure 12. Hydrostatic pressure contour for compression problem.

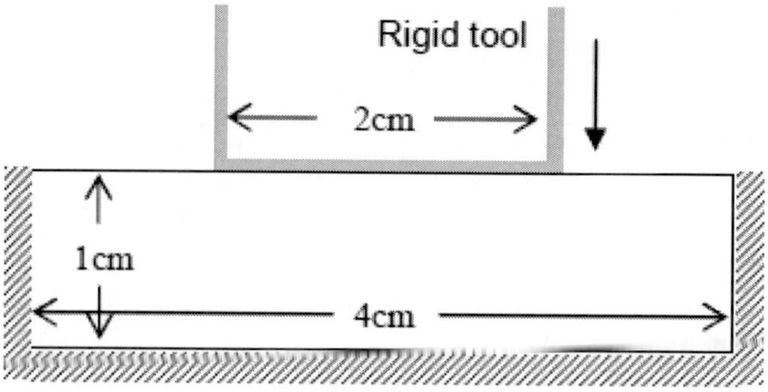

Figure 13. Punch problem.

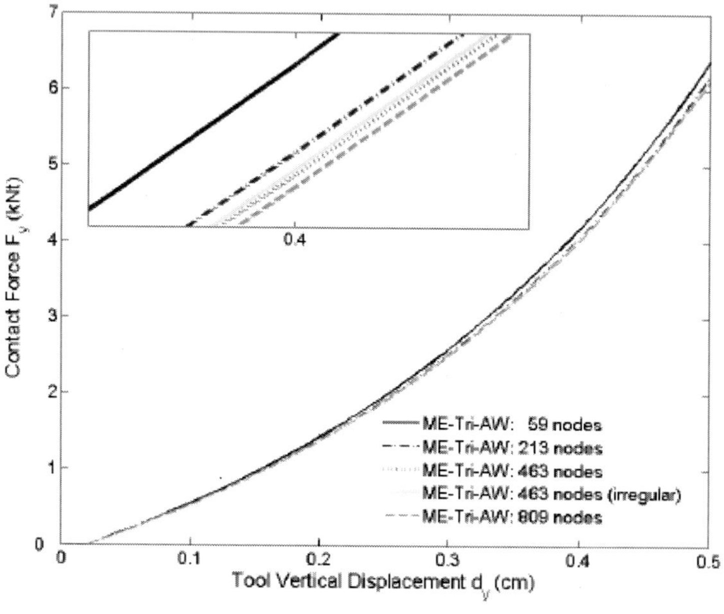

Figure 14. Convergence study of contact force using ME-Tri-AW.

The contact between the rigid tool and workpiece is assumed to be frictionless. The material properties are: Young's modulus $E = 2.0\text{GPa}$, Poisson's ratio $\nu = 0.3$, and an isotropic hardening rule is given by

$$\sigma_y\left(\bar{e}^p\right) = \sigma_y^0\left(1 + \alpha\bar{e}^p\right)^\beta \tag{54}$$

where is yield radius and is effective plastic strain, respectively, and $\sigma_y^0 = 0.02\text{GPa}$, $\alpha = 1.0\text{GPa}$, $\beta = 1.1$.

The load-displacement responses from ME-Tri-AW using five different mesh refinements are shown in Figure 14. These results depict the convergence of ME-Tri-AW as the mesh is refined.

Figure 15 shows the deformation and pressure contour of ME-Tri-AW results when the rigid tool reaches vertical displacement of $d_y = 0.35\text{cm}$. In comparison, Figure 16 plots the results of GMF-GI and NICE-T3 under the same tool displacement.

Table 1 reports the comparison of vertical contact forces in three methods using the discretization of 128 triangular elements. GMF-GI solution exhibits severe pressure oscillation as shown in Figure 16 (a). On the other hand, the solution from NICE-T3 is relatively softer and suffers from the spurious deformation mode as shown in Figure 16 (b). In contrast, the proposed ME-Tri-AW presents better deformation and smooth pressure field over the other two methods. The results of load-displacement curves in Figure 14 and the convergence study in Figure 15 also indicate that the ME-Tri-AW solutions are insensitive to the mesh irregularity.

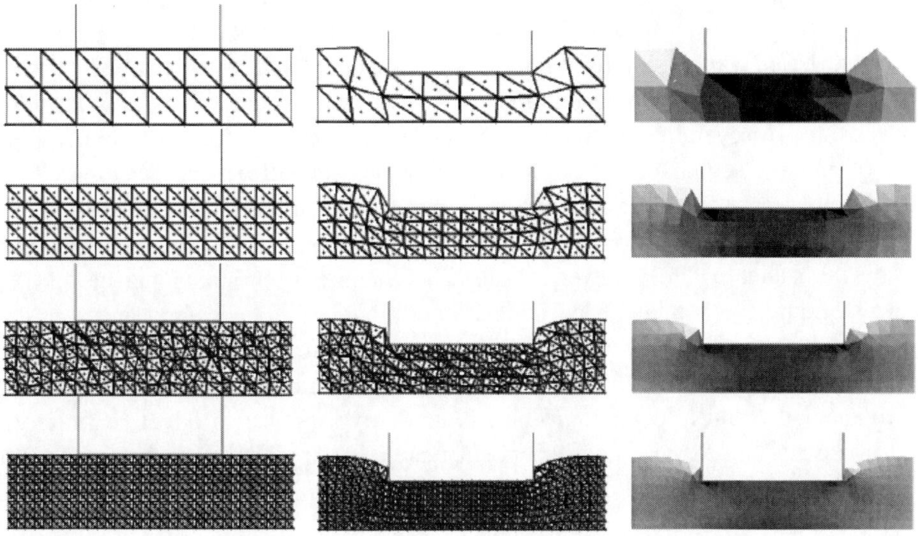

a b c

[a]discretization.
[b]deformation.
[c]hydrostatic pressure.

Figure 15. ME-Tri-AW results using four different discretizations ($d_y = 0.35\text{cm}$).

a b

[a]GMF-GI.
[b]NICE-T3.

Figure 16. Hydrostatic pressure results ($d_y = 0.35\text{cm}$).

Table 1. Vertical contact forces on the rigid tool ($d_y = 0.35\text{cm}$)

Tri elements	GMF-GI	NICE-T3	ME-Tri-AW
128	3275.64	2886.64	3229.31

CONCLUSION

We have presented a nonlinear version of the meshfree-enriched finite element method (ME-FEM) for the large deformation analysis of path-independent and path-dependent materials. We start with an introduction of smoothed deformation gradients constructed using a convex meshfree approximation [44] based on a meshfree-enriched finite element triangulation proposed in [45]. The introduced smoothed deformation gradient allows us to degenerate the three-fields Hu-Washizu-de Veubeke variational formulation to a displacement-based variational formulation that can be solved with relatively ease in the nonlinear analysis. The presented nonlinear formulation has demonstrated its accuracy and robustness through several benchmark examples and proven it is free of volumetric locking and free of pressure oscillation. In summary, the proposed nonlinear meshfree-enriched finite element formulation offers the following main attractions for the large deformation analysis: (1) its simplicity to be fitted into the conventional displacement-based finite element code and solved by the standard direct solver; (2) its convenience in enriching the meshfree nodes without redefining the boundary and contact conditions; (3) its accuracy and robustness in delivering volumetric locking-free and pressure oscillation-free solution in finite strain analysis; (4) its applicability to both path-independent and path-dependent materials. The proposed method could also be a promising alternative for use in adaptive finite element method involving large strain path-dependent materials. Application such as the r-adaptive finite element method in the three-dimensional metal forging and extrusion simulations will be considered in the near future.

ACKNOWLEDGMENTS

The authors would like to thank Dr. John O. Hallquist of LSTC for his support to this research. The support from Yokohama Rubber Co, Ltd, Japan is also gratefully acknowledged.

REFERENCES

[1] F. M. Andrade Pires, E. A. de Souza Neto, J. L. de la Cuesta Padilla, An assessment of the average nodal volume formulation for the analysis of nearly incompressible solids under finite strains, *Int. J. Numer. Methods Engrg.* 20 (2004) 569-583.

[2] D. N. Arnold, F. Brezzi, L. P. Franca, A stable finite element for the Stokes equations, *Calcolo* 21 (1984) 337-344.

[3] F. Auricchio, L. Beirao de Veiga, A. Buffam C. Lovadina, A. Reali, G. Sangalli, A fully locking-free isogeometric approach for plane linear elasticity problems: a stream function formulation, *Comput. Methods Appl. Mech. Engrg.* 197 (2007) 160-172.

[4] K. J. Bathe, *Finite element procedures*, Prentice-Hall, New Jersey, 1996.

[5] I. Babuska, The finite element method with Lagrangian multipliers, *Numer. Math.* 20 (1973) 179-192.

[6] I. Babuska, J. M. Melenk, The partition of unity method, *Int. J. Numer. Methods Engrg.* 40 (1997) 727-758.

[7] T. Belytschko, Y. Y. Lu, L. Gu, Element-free Galerkin methods, *Int. J. Numer. Methods Engrg.* 37 (1994) 229-256.

[8] T. Belytschko, W. K. Liu, B. Moran, *Nonlinear finite elements for continua and structures* (third ed.), John Wiley and Sons, New York, 2001.

[9] J. Bonet, A. J. Burton, A simple average nodal pressure tetrahedral element for incompressible and nearly incompressible dynamic explicit applications, *Commun. Numer. Methods Engrg.* 14 (1998) 437-449.

[10] L. Chen, G. R. Liu, N. Nourbakhsh-Nia, K. Zeng, A singular edge-based smoothed finite element method (ES-FEM) for biomaterial interface cracks, *Comput. Mech.* 45 (2010) 109-125.

[11] J. S. Chen, C. Pan, C. T., A pressure projection method for nearly incompressible rubber hyperelasticity, Part I: Theory, *J. Appl. Mech.* 63 (1996) 862-868.

[12] J. S. Chen, C. T. Wu, C. Pan, A pressure projection method for nearly incompressible rubber hyperelasticity, Part II: Applications, *J. Appl. Mech.* 63 (1996) 869-876.

[13] J. S. Chen, W. Han, C. T. Wu, W. Duan, On the perturbed Lagrangian formulation for nearly incompressible and incompressible hyperelasticity, *Comput. Methods Appl. Mech. Engrg.* 142 (1997) 335-351.

[14] J. S. Chen, S. Yoon, H. P. Wang, W. K. Liu, An improved reproducing kernel particle method for nearly incompressible finite elasticity, *Comput. Methods Appl. Mech. Engrg.* 181 (2000) 117-145.

[15] J. S. Chen, C. T. Wu, S. Yoon, Y. You, A stabilized conforming nodal integration for Galerkin mesh-free methods, *Int. J. Numer. Methods Engrg.* 50 (2001) 435-466.

[16] S. De, K. J. Bathe, Displacement/pressure mixed interpolation in the method of finite spheres, *Int. J. Numer. Methods Engrg.* 51 (2001) 275-292.

[17] J. Dolbow, T. Belytschko, Volumetric locking in the element free Galerkin method, *Int. J. Numer. Methods Engrg.* 46 (1999) 925-942.

[18] J. Dolbow, A. Devan, Enrichment of enhanced assumed strain approximations for representing strong discontinuities patch test, *Int. J. Numer. Methods Engrg.* 59 (2004) 47-67.

[19] C. A. Duarte, J. T. Oden. An h-p adaptive method using clouds, *Comput. Methods Appl. Mech. Engrg.* 39 (1996) 237-262.

[20] T. Elguedj, Y. Bazilevs, V. M. Calo, T. J. R. Hughes, $\overline{B}$ and projection methods for nearly incompressible linear and non-linear elasticity and plasticity using higher-order NURBS elements, *Comput. Methods Appl. Mech. Engrg.* 197 (2008) 2732-2762.

[21] Y. Guo, M. Ortiz, T. Belytschko, E. A. Repetto, Triangular composite finite elements, *Int. J. Numer. Methods Engrg.* 47 (2000) 287-316.

[22] P. Hauret, E. Kuhl, M. Ortiz, Diamond elements: a finite element/discrete-mechanics approximation scheme with guaranteed optimal convergence in incompressible elasticity, *Int. J. Numer. Methods Engrg*. 72 (2007) 253-294.

[23] Z. C. He, G. R. Liu, Z. H. Zhong, S. C. Wu, G. Y. Zhang, A. G. Cheng, An edge-based smoothed finite element method (ES-FEM) for analyzing three-dimensional acoustic problems, *Comput. Methods Appl. Mech. Engrg*. 199 (2009) 20-33.

[24] T. J. R. *Hughes, The finite element method*, Prentice-Hall: Englewood Cliffs, NJ, 2000.

[25] T. J. R. Hughes, J. A. Cottrell, Y. Bazilevs, Isogeometric analysis: CAD, finite elements, NURBS, exact geometry and mesh refinement, *Comput. Methods Appl. Mech. Engrg*. 194 (2005) 4135-4195.

[26] P. Krysl, B. Zhu, Locking-free continuum displacement finite elements with nodal integration, *Int. J. Numer. Methods Engrg*. 76 (2008) 1020-1043.

[27] B. P. Lamichhane, Inf-sup stable finite element pairs based on dual meshes and bases for nearly incompressible elasticity, *IMA J. Numer. Anal.* 29 (2009) 404-420.

[28] G. R. Liu, T. Nguyen-Thoi, *Smoothed finite element method*, CRC Press, Boca Raton, 2010.

[29] W. K. Liu, J. S.-J. Ong, R. A. Uras, Finite element stabilization matrices - a unification approach, *Comput. Methods Appl. Mech. Engrg*. 53 (1986) 13-46.

[30] W. K. Liu, S. Jun, S. Li, J. Adee, T. Belytschko, Reproducing kernel particle methods for structural dynamics, *Int. J. Numer. Methods Engrg*. 38 (1995) 1655-1679.

[31] D. S. Malkus, T. J. R. Hughes, Mixed finite element methods-reduced and selective integration techniques: a unification of concepts, *Comput. Methods Appl. Mech. Engrg*. 15 (1978) 63-81.

[32] A. Ortiz, M. A. Puso, N. Sukumar, Maximum-Entropy meshfree method for compressible and near-incompressible elasticity, *Comput. Methods Appl. Mech. Engrg*. 199 (2010) 1859-1871.

[33] C. K. Park, C. T. Wu, C. D. Kan, On the analysis of dispersion property and stable time step in meshfree method using the generalized meshfree approximation, *Finite Elem. Anal. Des*. 47 (2011) 683-697.

[34] R. W. Peen, Volume changes accompanying the extension of rubber, *Trans. Soc. Rheol*., 14 (1970) 509-517.

[35] M. A. Puso, J. Solberg, A stabilized nodally integrated tetrahedral, *Int. J. Numer. Methods Engrg*. 67 (2006) 841-867.

[36] R. S. Rivlin, Large elastic deformation of isotropic materials, Part VI, further results in the theory of torsion, shear and flexure, *Philos. Trans. R. Soc. of London*, A242 (1949) 173-195.

[37] J. C. Simo, T. Hughes, On the variational foundation of assumed strain methods, *ASME J. Appl. Mech*. 53 (1986) 51-54.

[38] K. R. Srinivasan, K. Matous, P. H. Geubelle. Generalized finite element method for modeling nearly incompressible bimaterial hyperelastic solids, *Comput. Methods Appl. Mech. Engrg*. 197 (2008) 4882-4893.

[39] R. Stenberg, Error analysis of some finite element methods for the Stokes problem, *Math. Comput*. 54 (1990) 495-508.

[40] A. C. Stevenson, Some boundary problems of two-dimensional elasticity, *Philos. Mag*. 34 (1943) 766-793.

[41] Y. Vidal, P. Villon, A. Huerta, Locking in the incompressible limit: pseudo-divergence-free element free Galerkin, *Commun. Numer. Methods Engrg*. 19 (2003) 725-735.

[42] K. Washizu, *Variational Methods in Elasticity and Plasticity* (3rd edn.), Pergamon Press, New York, 1982.

[43] C. T. Wu, M. Koishi, A meshfree procedure for the microscopic analysis of particle-reinforced rubber compounds, *Interaction and Multiscale Mechanics* 2 (2009)147-169.

[44] C. T. Wu, C. K. Park, J. S. Chen, A generalized meshfree approximation for the meshfree analysis of solids, *Int. J. Numer. Methods Engrg*. 85 (2011) 693-722.

[45] C. T. Wu, W. Hu, Meshfree enriched simplex elements with strain smoothing for the finite element analysis of compressible and nearly incompressible solids, *Comput. Methods Appl. Mech. Engrg*. 200 (2011) 2991-3010.

[46] O. C. Zienkiewicz, R. L. Taylor, *The finite element method* (third ed.), McGraw-Hill, London, 1987.

In: Focus on Nonlinear Analysis Research
Editors: Gualtiero Padovani and Manuel Occhino
ISBN: 978-1-62417-350-9
© 2013 Nova Science Publishers, Inc.

Chapter 6

NONLINEAR ANALYSIS RESEARCH IN BIOMEDICAL ENGINEERING

***Paolo Melillo*[1,2,*], *Leandro Pecchia*[3] and *Mauro Ursino*[1]**

[1] Department of Electronics, Computer Science and Systems, University of Bologna, Bologna, Italy
[2] Department of Ophthalmology, Second University of Naples, Naples, Italy
[3] Department of Electrical and Electronic Engineering, University of Nottingham, Nottingham, United Kingdom

ABSTRACT

Biological phenomena have shown behaviors that are strongly nonlinear. Therefore, in order to correctly enhance information from biomedical signals, analysis based on nonlinear methods should be adopted beyond the classical linear modeling. These approaches reflect better the nonlinear nature of biological phenomena, and are more suitable to model and interpret information of modulation phenomena, on-off mechanism of response, variable gain, threshold and saturation characteristics, and even chaotic behaviors. In this chapter, we will review nonlinear methods applied for the processing of biomedical signals.

We will refer to the Heart Rate Variability, the series of heartbeats, as this signal is commonly analyzed using both linear and nonlinear methods. Nonetheless, these methods have been widely applied to the study of other biomedical signals. The methods reviewed in this chapter are the Poincaré plot, the fractal dimension, the correlation dimension, the detrended fluctuation analysis, the approximate entropy, the sample entropy, and the recurrence plot. We will show the potentiality of nonlinear methods compared to the linear modeling with reference to published studied.

---

* Corresponding author: Paolo Melillo. Email: paolo.melillo2@unibo.it.

INTRODUCTION

In the investigation of biological phenomena, biological systems have been shown to exhibit typical nonlinear characteristics, such as bio-potentials generated by pacemaker cells in heart during ventricular tachycardia, or abrupt changes like the sudden onset of seizure by brain neurons. Indeed, the two central features of linear systems, proportionality and superposition, are hardly satisfied by biological systems. In fact, in biomedical systems, small changes in the inputs can cause not-proportionally huge changes in the output and the composition of multiple inputs (in time and in space) could not be explained by analyzing individually the answers of the system to the single inputs. In several different biomedical engineering research activities, many nonlinear behaviors have been observed, such as hysteresis, nonlinear oscillations, fractal behavior, and high dependence from parameters. These behaviors require the adoption of nonlinear algorithms eventually with some modifications.

When applying nonlinear methods on biomedical signal processing, the following issues should be considered and dealt with:

- low signal/noise ratio: high level of random noise in the biosignal and low intensity of sources require the adoption of methods robust to noise influence;
- the signal sources are not human-designed: this means that the extraction of the information is unsupervised;
- low repeatability: this means small datasets and short recordings, which may result in inaccurate estimation of parameters and inaccurate estimation of high frequency spectrum estimation;
- non-stationarity: this does not allow the use of Fourier transform to estimate the spectrum, requiring pre-processing of raw data before analysis [1] or the adoption of other periodograms.
- low output resistance of the bio-sources: thus the measures alter the signal shape;
- low variability of signals: all the bio-potential have all a low and narrow bandwidths, this create many problems (i.e. electrodes polarization) and the co-esistence of noises and disturbances in the same band (limited effect of filters, differential input required).

In this chapter, we present the main nonlinear methods applied for processing of biomedical signals, focusing on the analysis of the Heart Rate Variability (HRV), that is, the variation over time of the period between consecutive heartbeats (RR intervals)[2]. We chose this biomedical signal since HRV is a widely studied time series investigated using both linear and nonlinear methods.

Linear methods, and consequently linear measures, of HRV have been standardized in international guidelines since 1996[2]. Since then, many nonlinear methods have been introduced in physiology (and also adapted) with reference to heart beat analysis, and the nonlinear features, in particular complexity, have shown to be able to distinguish physiological conditions from pathological ones better than linear measures.

An example is the higher complexity in the human heartbeat which is observed under resting or relatively inactive conditions in healthy subjects compared to cardiac patients, particularly, suffering from severe congestive heart failure. This free spontaneous complex variability is hypothesized to be an indicator of complexity reserve, which can be utilized when the individual is asked to perform a task, either mental or physical, producing a specific response. In patients, heart rate complexity is reduced and the complexity values, even at rest, are similar to those of healthy individuals at the brink of exhaustion. For these patients, the number of different dynamical states is diminished in comparison to those available in healthy subjects. Therefore, the difference in heart rate complexity between resting and active conditions is smaller for congestive heart failure patients than for healthy subjects. In other words, normal subjects lay in a dynamic non-stable equilibrium state, which allows fast changes to react to environmental stimuli.

NONLINEAR METHODS IN BIOLOGY AND PHYSIOLOGY

In this chapter, we will briefly describe the general use of nonlinear methods with reference to a generic time-series x_1, x_2, …, x_N,, as appropriate, and to the RR interval time series RR_1, RR_2, …, RR_N. The most common nonlinear methods applied to HRV analysis are the following ones: poincaré plot [3, 4], approximate entropy[5], sample entropy[5], fractal dimension, correlation dimension[6], detrended fluctuation analysis[7, 8], and recurrence plot [9-11].

The pathophysiological condition considered in the current chapter are Congestive Heart Failure (CHF), acute myocardial infarction (AMI), supraventricular arrhythmia (SA), atrial fibrillation (AF), ventricular fibrillation (VF), sick sinus syndrome (SSS).

Among the several linear features previously applied to HRV analysis, both in the time and in the frequency domains, we remember, for the purpose of this chapter, the standard deviation of the successive intervals, further referred as *SDRR* or *SDNN* if the beats have a physiologic shape, and the standard deviation of the successive differences of the RR intervals, further referred as *SDSD*.

POINCARE PLOT

The Poincaré Plot (PP), also known as return map, is a common graphical representation in which each point is represented as a function of the previous one, that is a plot of (x_1,x_2), (x_2,x_3),…,(x_{N-1},x_N).

In HRV analysis, PP is the scatterplot of the successive RR versus previous one. Figure 1 shows an example. A widely used approach to analyse the Poincaré plot of RR series consists in fitting an ellipse oriented according to the line-of-identity and computing the standard deviation of the points perpendicular to and along the line-of-identity referred as SD_1 and SD_2, respectively[3].

It has been shown that SD_1 and SD_2 are related to linear measures of HRV[3]. In particular, SD_1 is equivalent to the *SDRR*, except that it is scaled by $1/\sqrt{2}$, while SD_2 is related to existing linear HRV measures as described by the equation 2.

$$SD_1 = \frac{SDSD}{\sqrt{2}} \quad (1)$$

$$SD_2 = \sqrt{2SDRR^2 - \frac{1}{2}SDSD^2} \quad (2)$$

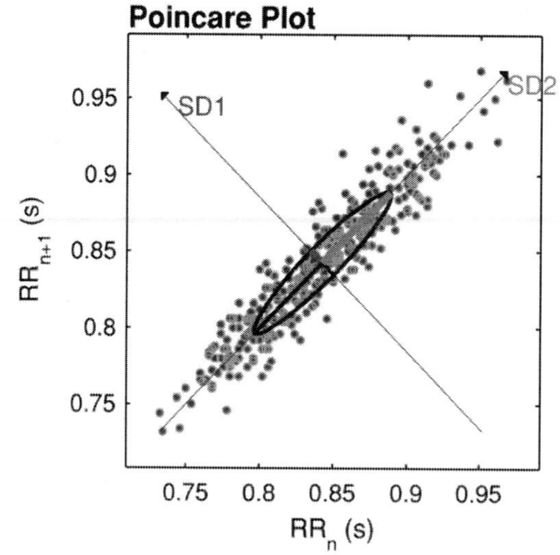

Figure 1. Example of Pointcarè plot of RR time series.

Moreover, the two PP measures are related to the autocovariance function Φ_{RR} as described in the following equations:

$$SD_1 = \sqrt{{\Phi_{RR}}^2(0) - {\Phi_{RR}}^2(1)} \quad (3)$$

$$SD_2 = \sqrt{{\Phi_{RR}}^2(0) + {\Phi_{RR}}^2(1)} \quad (4)$$

For that reason, these two PP measures could not provide independent nonlinear information and in recent review on HRV[12] they are discusses in the standard time domain analysis instead of nonlinear methods.

Two kind of generalization of the PP are proposed in the literature[3]: the lagged PP and the higher order PP. The lagged PP is the plot of RR_{n+m} against RR_n where m is chosen from 2 to some small positive value (not higher than 8). SD_1 and SD_2 are computed similarly as lag m set to 1 and are also related to the autocovariance function as reported in the following equations:

$$SD_1(m) = \sqrt{{\Phi_{RR}}^2(0) - {\Phi_{RR}}^2(m)} \quad (5)$$

$$SD_2(m) = \sqrt{{\Phi_{RR}}^2(0) + {\Phi_{RR}}^2(m)} \quad (6)$$

For that reason, the set of lagged PP is a description of the autocovariance function.

PP of order m is a m-dimensional scatter-plot of the m-ples (RR_n, RR_{n+1},..., RR_{n+m}). This plot resulted in 2-dimensional projection into each of the coordinate planes (RR_n, RR_{n+1}), (RR_{n+1}, RR_{n+2}) ,...,(RR_n, RR_{n+m}). The first two projections are equivalent to standard PP, the last one is equivalent to a PP with lag m, and the other projections are equivalent to PP with lag up to m. In other words, an order m PP is geometrically described by a set of PP with lag up to $m+1$.

Others techniques, such as central tendency measure[13], density based approach[14], have been applied in order to extract independent nonlinear information from PP, but they are not as widely used as SD_1 and SD_2.

Geometrically, SD_1 measures the width of the Poincaré cloud and, therefore, indicates the level of short-term HRV, while SD_2 measures the length of the cloud along the line-of-identity, reflecting the long-term HRV. Mathematically, equations 1 and equation 2 express these interpretation as SD_1 is a scaled SDSD, a measure of short-term (over each beat) variability, while SD_2 is the difference between total variability ($SDNN^2$) and short-term variability ($SDSD^2$). This could be explained by considering a time series which shows variability only over a single beat, such as a sequence alternating between two value (RR_I, RR_{II}, RR_I, RR_{II},...,RR_I, RR_{II}). The PP of this series shows a zero length and zero value of SD_2, which are coherent with the absence of long-term variability, while the SD_1 has a non-zero value, reflecting the short-term variability of the time series.

Several studies analysed PP measures of HRV, among them Isler et al.[15] proposed automatic classifier for CHF detection and the two combinations of features achieving the best performance included PP measures. Moreover, a reduction of SD_1 was observed in stressful mental situations, such as precompetitive anxiety situations in swimmers[16], visual working memory tasks[17], and university examination[18].

APPROXIMATE ENTROPY

Approximate entropy ($ApEn$) is widely used method to measure the complexity of signal[19]. It shows the probability that similar observation patterns do not repeat. If a time series demonstrates complex, irregular behaviour, than it will have a high $ApEn$ values. For instance, sinusoids give approximately zero value of $ApEn$. $ApEn$ showed several advantages: it can be applied for both short-term and long-term recordings, it is scale invariant, model independent, easy to use and it is able to discriminate time series for which clear future recognition is difficult[12, 20]. For that reason, it has been applied in different fields, particularly in cardiovascular signal analysis to assess the irregularity of the RR series[5].

The algorithm for the computation of $ApEn$ of an RR time series was described here.

A series of vector of length m X_1,X_2, ..., X_{N-m+1} is constructed from the RR intervals as follows:

$$X_i,=[RR_i, RR_{i+1} \ldots RR_{i+m-1}].$$

The distance between vectors and is defined as the maximum absolute difference between their respective scalar components. For each vector X_i, the relative number of vectors for which $d[X_i, X_j] \le r$, $C_i^m(r)$ is computed where r is referred as a tolerance value (see equation 7).

$$C_i^m(r) = \frac{number\ of \{d[X_i, X_j] \le r\}}{N-m+1} \quad \forall j \tag{7}$$

Then, the following index is computing by taking natural logarithm of each and averaging them over i.

$$\Phi^m(r) = \frac{1}{N-m+1} \sum_{i=1}^{N.-m+1} \ln C_i^m(r) \tag{8}$$

Finally, the approximate entropy is computed as:

$$ApEn(m, r, N) = \Phi^m(r) - \Phi^{m+1}(r) \tag{9}$$

The value of m, embedding dimension, and of r are required to be specified a priori. Several clinical studies[19, 21, 22] have shown that either m=1 or 2 and r between 0.1 and 0.2 times the SDRR are suitable to provide valid value of *ApEn*. However, a recent study[20] recommended the use of the r value (r_{max}) which maximizes the *ApEn* ($ApEn_{max}$). This conclusion was derived by the observation that the *ApEn* computed with value of r within the recommended range 0.1 - 0.2 provided misleading results in simulated signals and by the results of a previous study on human HRV data[23]. For example, a selection of r=0.25 resulted in a 12% decrease of *ApEn* values, whereas r=0.1 results in 9% increase as subjects changed their body position from supine to upright.

In order to avoid the calculation of *ApEn* for each possible r value to find the maximum value, nonlinear models have been proposed and validated to estimate r_{max} value from variability of the signals. In particular, for m=2, Chon et al. [20] proposed the following formula:

$$r_{chon} = (-0.036 + 0.26\sqrt{SDDS / SDNN}) / \sqrt[4]{N/1000} \tag{10}$$

where N denotes the length of the RR sequence.

A recent study by Liu[24] aimed to verify whether Chon's method was appropriate for HRV by comparing *ApEn* in two groups: healthy subject and patients suffering from CHF. *ApEn* value were computed with three different value of the threshold r :

$r = 0.2 * SDNN$ ($ApEn_{0.2}$);

$r = r_{max}$;

$r = r_{chon}$ ($ApEn_{Chon}$).

Surprisingly, only $ApEn_{Chon}$ (not $ApEn_{0.2}$ nor $ApEn_{max}$) was statistically different between the two groups.

Another recent study [25] analysed the three different type of *ApEn* in healthy subjects under stress compared to controlled resting condition. Also in this study, $ApEn_{max}$ was not statistically different between the two groups, while statistically significant differences were observed in $ApEn_{0.2}$ e $ApEn_{Chon}$. The findings of these studies could be explained according to the analysis of the Chon's empirical formula, particularly, its relationship with the ratio between short- and long- term variability. Further studies will focus on development of methods which can reduce the influence from the different threshold values *r* in *ApEn* computation.

As in some pathophysiological conditions, the heart rate times became more regular with repetition of similar patterns, lower values of *ApEn* were observed in several clinical conditions, SA[26], AMI patients[27], CHF patients[26], in non survivors among acute brain injury patients[28].

SAMPLE ENTROPY

Sample Entropy (*SampEn*) is a relatively new feature introduced by Richman et al.[5] to measure the complexity and the regularity of clinical time series. It is very similar to *ApEn*, with two important differences in its calculation: first, in the computation of the comparison of the vector X(i) with itself is included in the count for *ApEn* (self-match problem), while this comparison is excluded for *SampEn*; secondly, the logarithm is applied instead of subtraction in the final step. These changes aims to remove the bias in *ApEn*, as the count of the self-comparison in *ApEn* lower its value and the signals are interpreted as more regular than they are.

Formally, the three steps of *SampEn* computation are described by the following formulae:

$$C_i^m(r) = \frac{number\ of\{d[X_i, X_j] \le r\}}{N-m+1} \quad \forall j \ne i \tag{11}$$

$$\Phi^m(r) = \frac{1}{N-m+1} \sum_{i=1}^{N-m+1} \ln C_i^m(r) \tag{12}$$

$$SampEn(m,r,N) = \log \frac{\Phi^m(r)}{\Phi^{m+1}(r)} \tag{13}$$

SampEn, in theory, does not depend on N, but it also relies on the choice of the parameters m and r, such as *ApEn*. However, the dependence on the parameter r is different: *SampEn* decreases monotonically when r increases. With high value of N and r, *SampEn* and

ApEn provide comparable results[12], for that reason, the applications of the two measures are very similar.

FRACTAL DIMENSION

The term "fractal" was first defined as a geometric concept[29], referring to an object satisfying two properties: self-similarity and fractional dimensionality. The former property means that an object is composed of subunits (and sub-subunits on multiple levels) that statically resemble the structure of the whole object. The latter property means that the object has a fractional dimension.

For example, in bidimensional curve, to verify self-similarity, a subset of the object is rescaled to the same size of the original object, using the same magnification factor for both its width and height, and its statistical properties are compared with those of the original object. Mathematically, this property should hold on all scales, however, in the real world, there are necessarily lower and upper bounds over which such self-similar behavior applies. Moreover, the strict criterion requires that all the statistical properties (including all higher moments) are identical.

Therefore, in practice, a weaker criterion is adopted by examining only the means and variances (first and second moments) of the distribution. The second criteria distinguishes fractal from Euclidean objects, which have integer dimension. For example, a square satisfies self-similarity as it can be divided into smaller subunits that resemble the large square, but it is not a fractal since it has an integer (2) dimension. Fractal-like appearance was observed in several cardiovascular structures, such as the arterial and venous tree and His-Purkinje network.

The concept of fractal has been extended to the analysis of time series. For instance, in order to verify self-similarity in time series, a subset of the time series is selected and rescaled, but with two different magnification factors (since the two axes have independent physical units). Mathematically, a time series is self-similar if

$$x(t) \equiv a^{a} x\left(\frac{t}{a}\right) \tag{14}$$

where a is the self-similarity parameter or scaling exponent and the operator $\equiv$ indicate that the statistical properties of both side of the equation are identical; however, as already stated, in practice, only the first and second moment statistical properties are compared. The scaling exponent could be estimated by Detrended Fluctuation Analysis as discussed in the following paragraph, while we describe here the Higuchi's algorithm for estimation of the fractal dimension (FD) [30].

First, a k time series was constructed as follows:

$$x_m^k = [RR(m), RR(m+2k), \ldots, RR(m+[n-mk]k)] \; with \; i = 1,2, \ldots., k \tag{15}$$

where m is the initial time value, and k the discrete time intervals.

For each time series, the length is defined as follows:

$$L_m(k) = \frac{1}{\lfloor \frac{N-m}{k} \rfloor k} \sum_{i=1}^{\lfloor \frac{N-m}{k} \rfloor} |x(m+ik) - x(m+(i-1)k)| \tag{16}$$

where N is the total length of the time series, the $\lfloor \quad \rfloor$ indicates the Gaussian operator and $|x|$ indicates the module of x. The estimate of the FD according to Higuchi is the slope, computed by least-squares linear best fit, of the curve $\ln[L_m(k)]$ versus $ln(1/k)$.

CORRELATION DIMENSION

The correlation dimension *CD* is one of the most widely used measures of the fractal dimension and has been adopted to measure the complexity for the HRV time series[6]. As for *ApEn* computation, the reconstruction of the attractor is the first step, that is, a series of vector of length m $X_1, X_2, \ldots, X_{N-m+1}$ is constructed from the RR intervals as follows:

$$X_i, = [RR_i, RR_{i+\tau} \ldots RR_{i+\tau(m-1)}] \tag{17}$$

where τ is the time delay and m is the embedding dimension. The second step is the estimation of Euclidean distances between each couple of vectors:

$$d[X_i, X_j] = \sqrt{\sum_{k=1}^{m} (X_i(k) - X_j(k))^2} \tag{18}$$

Then, the idea is to construct a function which estimates the probability that two arbitrary points on the orbit are close than r. So, the correlation integral function is determined by the following formula

$$C_m(r) = \frac{1}{N_m(N_m - 1)} \sum_{i}^{N_m} \sum_{j=1}^{N_m} \vartheta(r - d[X_i, X_j]) \tag{19}$$

where and ϑ is the Heavide function, that is, $\vartheta(x) = 0$ if $x < 0$, otherwise $\vartheta(x) = 1$.

The correlation dimension is defined as the following limit value:

$$CD(m) = \lim_{r \to 0} \lim_{N \to \infty} \frac{\log C^m(r)}{\log r} \tag{20}$$

In practice, this limit value is approximated by the slope of the regression curve ($\log r$, $\log C^m(r)$).

The appropriate value of τ could be chosen using the minimal mutual information technique [31, 32], while the value of m could be estimated with the methods proposed by Grossberger and Procaccia[33]; in HRV analysis the values of 1 and 10 are widely used value for τ and m, respectively. The CD value is higher for the chaotic data and it decreases as the variation in RR time series becomes less chaotic or rhythmic. Higher value of CD were observed in normal sinus rhythm ECG compared to different arrhythmic ECG [34, 35].

DETRENDED FLUCTUATION ANALYSIS

Detrended Fluctuation Analysis is used to quantify the fractal scaling properties of short RR time series, by measuring the correlation within the signal [7, 8]. It is a modified version of root-mean-square analysis of random walks [36] and consists into: integration of the root-mean-square fluctuation, computation of detrended time series at observation window of different size and plot against the windows sizes on a log-log scale.

Formally, the following steps are performed in DFA

1) The average of the RR interval series is calculated on all the N samples. The alternate component of RR interval series, which is defined as RR minus its average value $\overline{RR}$, is integrated:

$$(21)\ y(k)=\sum_{j=1}^{k}(RR_j-\overline{RR}),\ k=1,....,N\ .$$

2) The integrated series is divided into non-overlapping segments of equal length n. A least square line is fitted within each segment, representing the local trends with a broken line. This broken line is referred as $y_n(k)$, where n denotes the length of each segment.
3) The integrated time series is detrended as follows: $y(k)-y_n(k)$. The root-mean-square fluctuation of the detrended time series is computed according to the following formula:

$$F(n)=\sqrt{\frac{1}{N}\sum_{k=1}^{N}(y(k)-y_n(k))^2}\ . \qquad (22)$$

4) The steps from 2 to 4 are repeated for n from 4 to 64.
5) Representing the function $F(n)$ in a log-log diagram, a linear relationship means presence of fractal scaling and the scaling exponent α is computed the slope of the regression line. The value of α for white noise, for fractal-like signal and Brownian noise (integral of Gaussian noise) are 0.5, 1 and 1.5, respectively[22, 36].

In normal subjects α is close to one, while lower values were observed in very highly varying RR time series, such as AF and VF rhythms, but in rhythmically varying RR time series, such SSS, and ischemic or dilated cardiomyopathy, the values are higher [34, 35]. In HRV analysis, other two additional indexes are usually computed [37, 38]: short-term fluctuations (α_1) as the slope of the regression line relating to with n within 4-16; long-term fluctuations (α_2) as the slope of the regression line relating to with n within 16-64.

RECURRENCE PLOT

Recurrence Plot (RP) was introduced by Eckmann et al.[39] as a graphical tool to discover hidden periodicities, difficult to be detected otherwise, and can be used to reveal non-stationarity in the time series. RP is designed according to the following steps.

As in CD computation, vectors $X_i = (RR_i, RR_{i+\tau}, ..., RR_{i+(m-1)\tau})$, with $i=1,..., K$, with $K=[N-(m-1)\tau]$, where m is the embedding dimension and τ is the embedding lag, are defined.

The RP is a K-dimensional matrix of dots, where a dot is placed if the Euclidean distance between X_i and X_j is lower than a threshold value r.

Formally, the following steps are suggested for achieving the RP:

(1) A K-dimensional square matrix M_1 is calculated computing the Euclidean distances of each vector X_i from all the others.
(2) A K-dimensional square matrix M_2 is calculated as the matrix whose elements $M_2(i,j)$ are defined as:

$$M_2(i,j) = \begin{cases} 1 & if \quad M_1(i,j) < r \\ 0 & if \quad M_1(i,j) > r \end{cases} \tag{23}$$

The RP is the representation of the matrix M_2 in which a dot is associated to one value, that is, an image in which black pixels correspond to ones and white pixels to zeros.

M_1 is a symmetrical matrix as the distance between X_i and X_j is equal to the one between X_j and X_i and consequently, RP is a symmetric image along the diagonal.

According to findings by Niskander et al. and Dabire et al. [21, 40], the following values of the parameters should be chosen: $m = 10; \tau = 1; r = \sqrt{m} * SDRR$.

In the RP, lines are defined as series of diagonally adjacent black points with no white space. The length l of a line is the number of points which the line consists of.

RP of HRV in normal healthy subject has diagonal lines and less squares reflecting high variation in heart rate, while abnormal recordings, such as from ischemic or dilated cardiomyopathy, resulted in a RP with more squares reflecting periodicity and lower variation in heart rate.

Moreover, some measures of RP are widely computed: recurrence rate *(REC)* defined in equation 24; maximal length of lines (l_{max}); mean length of lines (l_{mean}); the determinism (*DET*) defined in equation 25; the Shannon Entropy (*ShEn*) defined in equation 26.

$$REC = \frac{1}{K^2}\sum_{i=1}^{K}\sum_{j=1}^{K} M_2(i,j) \tag{24}$$

$$DET = \frac{\sum_{l=2}^{l_{max}} l * N_l}{\sum_{i=1}^{K}\sum_{j=1}^{K} M_2(i,j)}, \tag{25}$$

with N_l = number of lines of length l

$$ShEn = \sum_{l=l_{min}}^{l_{max}} n_l * \ln n_l, \tag{26}$$

with n_l = percentage of N_l over all the number of lines.

RP was used to detect life treating arrhythmias like ventricular tachycardias by Marwan et al.[42].

Requirements for Nonlinear Analysis

Before using the previous indexes, nonlinearity should be checked and the *surrogate data test* is one of the most utilized tests for nonlinearity. It consists of statistical comparison of topological properties from the original time series with those from a surrogate time series with similar spectral properties. In particular, Theiler, which introduced this technique in 1992 [41], adopted the correlation dimension as topological property and proposed two different algorithms.

The first one (random phase) is based on the null hypothesis that the time series come from a linear stochastic process, and the surrogate data computed in order to have the same Fourier spectra as the raw data are generated by the following steps:

1) Computation of the Discrete Fourier Transform of the original time series
2) Randomization of the phases. The phases should also be symmetrized to obtain a real surrogate time series.
3) Inversion of the Discrete Fourier Transform.

The second one (data suffling) is based on the null hypothesis that the observed time series is a nonlinear static transformation of a linear stochastic process. In this algorithm, the value of the original time series are rescaled so that they become Gaussian, then the first algorithm is applied to the rescaled original.

Finally, the surrogate time series is rescaled to have the same values as the original one. The advantage of the second algorithm compared to the first one is that the surrogate time series has the same amplitude distribution as it is a suffling of the original time series.

The selected topological property of surrogate data (consisting into several surrogate time series) is compared with the value of original time series by a statistical test (t-test with a significance level of 5%) [43] and if the null hypothesis is rejected the original time series is likely to include nonlinear features.

CONCLUSION

Biological systems have been shown to exhibit nonlinear behaviors, which cannot be exhaustively modeled using linear methods and measures. Thus, nonlinear methods have been adopted beyond the classical linear modeling in order to enhance information from biomedical signal processing. These approaches better reflect the nonlinear nature of biological phenomena, by capturing information of modulation phenomena, on-off mechanism of response, variable gain, threshold and saturation characteristics and chaotic behavior.

However, the biological systems have peculiar features that, in some cases, require modification in the standard nonlinear algorithms. When applying nonlinear methods on biosignals, the following issues should be considered and dealt with: high level of random noise in the biosignal; small dataset and short recordings; and nonstationarity.

The nonlinear methods presented in the current chapter are the Poincaré plot, the approximate entropy, the sample entropy, the fractal dimension, the correlation dimension analysis, the detrended fluctuation analysis, and the recurrence plot.

The algorithms are described with reference to heart beat time series, as it is a widely investigated time series in nonlinear domain, and the requirements for using nonlinear indexes (for instance nonlinearity test) are discussed.

Finally, in previous published studies these algorithms have been shown to provide reliable indexes for investigating and/or detecting pathophysiological conditions, such different arrhythmia (compared to normal sinus rhythm), CHF patients (compared to healthy subjects), mental stress conditions (compared to rest conditions).

ACKNOWLEDGMENTS

PM acknowledges support of this work through the Smart Health and Artificial intelligence for Risk Estimination (SHARE) project financed by Italian Ministry of Education, University and Scientific Research (Grant PON04a3_00139). LP acknowledges support of this work through the MATCH Program (EPSRC Grant EP/F063822/1).

REFERENCES

[1] Z. Chen, P. C. Ivanov, K. Hu *et al.*, “Effect of nonstationarities on detrended fluctuation analysis,” *Physical Review E,* vol. 65, no. 4, pp. 041107, 2002.

[2] M. Malik, J. T. Bigger, A. J. Camm *et al.*, "Heart rate variability: Standards of measurement, physiological interpretation, and clinical use," *Eur Heart J,* vol. 17, no. 3, pp. 354-381, March 1, 1996, 1996.

[3] M. Brennan, M. Palaniswami, and P. Kamen, "Do existing measures of Poincare plot geometry reflect nonlinear features of heart rate variability?," *Ieee Transactions on Biomedical Engineering,* vol. 48, no. 11, pp. 1342-1347, Nov, 2001.

[4] P. Melillo, R. Fusco, M. Sansone *et al.*, "Discrimination power of long-term heart rate variability measures for chronic heart failure detection," *Medical and Biological Engineering and Computing,* vol. 49, no. 1, pp. 67-74, 2011.

[5] J. S. Richman, and J. R. Moorman, "Physiological time-series analysis using approximate entropy and sample entropy," *American Journal of Physiology-Heart and Circulatory Physiology,* vol. 278, no. 6, pp. H2039-H2049, Jun, 2000.

[6] R. Carvajal, N. Wessel, M. Vallverdú *et al.*, "Correlation dimension analysis of heart rate variability in patients with dilated cardiomyopathy," *Computer Methods and Programs in Biomedicine,* vol. 78, no. 2, pp. 133-140, 2005.

[7] T. Penzel, J. W. Kantelhardt, L. Grote *et al.*, "Comparison of detrended fluctuation analysis and spectral analysis for heart rate variability in sleep and sleep apnea," *Ieee Transactions on Biomedical Engineering,* vol. 50, no. 10, pp. 1143-1151, Oct, 2003.

[8] C. K. Peng, S. Havlin, H. E. Stanley *et al.*, "Quantification of Scaling Exponents and Crossover Phenomena in Nonstationary Heartbeat Time-Series," *Chaos,* vol. 5, no. 1, pp. 82-87, Mar, 1995.

[9] L. L. Trulla, A. Giuliani, J. P. Zbilut *et al.*, "Recurrence quantification analysis of the logistic equation with transients," *Physics Letters A,* vol. 223, no. 4, pp. 255-260, Dec 9, 1996.

[10] C. L. Webber, and J. P. Zbilut, "Dynamical Assessment of Physiological Systems and States Using Recurrence Plot Strategies," *Journal of Applied Physiology,* vol. 76, no. 2, pp. 965-973, Feb, 1994.

[11] J. P. Zbilut, N. Thomasson, and C. L. Webber, "Recurrence quantification analysis as a tool for nonlinear exploration of nonstationary cardiac signals," *Medical Engineering and Physics,* vol. 24, no. 1, pp. 53-60, Jan, 2002.

[12] U. Rajendra Acharya, K. Paul Joseph, N. Kannathal *et al.*, "Heart rate variability: a review," *Med Biol Eng Comput,* vol. 44, no. 12, pp. 1031-51, Dec, 2006.

[13] K. Hnatkova, X. Copie, A. Staunton *et al.*, "Numeric processing of Lorenz plots of R-R intervals from long-term ECGs. Comparison with time-domain measures of heart rate variability for risk stratification after myocardial infarction," *J Electrocardiol,* vol. 28 Suppl, pp. 74-80, 1995.

[14] M. E. Cohen, D. L. Hudson, and P. C. Deedwania, "Applying continuous chaotic modeling to cardiac signal analysis," *Engineering in Medicine and Biology Magazine, IEEE,* vol. 15, no. 5, pp. 97-102, 1996.

[15] Y. Isler, and M. Kuntalp, "Combining classical HRV indices with wavelet entropy measures improves to performance in diagnosing congestive heart failure," *Computers in Biology and Medicine,* vol. 37, no. 10, pp. 1502-1510, Oct, 2007.

[16] J. C. Cervantes Blasquez, G. Rodas Font, and L. Capdevila Ortis, "Heart-rate variability and precompetitive anxiety in swimmers," *Psicothema,* vol. 21, no. 4, pp. 531-6, Nov, 2009.

[17] S. Mukherjee, R. Yadav, I. Yung *et al.*, "Sensitivity to mental effort and test-retest reliability of heart rate variability measures in healthy seniors," *Clinical Neurophysiology,* vol. 122, no. 10, pp. 2059-2066, Oct, 2011.
[18] P. Melillo, M. Bracale, and L. Pecchia, "Nonlinear Heart Rate Variability features for real-life stress detection. Case study: students under stress due to university examination," *Biomed Eng Online.,* vol. 10, no. 1, pp. 96., Nov 7, 2011.
[19] S. M. Pincus, "Approximate Entropy as a Measure of System-Complexity," *Proceedings of the National Academy of Sciences of the United States of America,* vol. 88, no. 6, pp. 2297-2301, Mar, 1991.
[20] K. H. Chon, C. G. Scully, and S. Lu, "Approximate Entropy for all Signals Is the Recommended Threshold Value r Appropriate?," *Ieee Engineering in Medicine and Biology Magazine,* vol. 28, no. 6, pp. 18-23, Nov-Dec, 2009.
[21] J.-P. Niskanen, M. P. Tarvainen, P. O. Ranta-aho *et al.*, "Software for advanced HRV analysis," *Computer Methods and Programs in Biomedicine,* vol. 76, no. 1, pp. 73-81, 2004.
[22] K. K. Ho, G. B. Moody, C. K. Peng *et al.*, "Predicting survival in heart failure case and control subjects by use of fully automated methods for deriving nonlinear and conventional indices of heart rate dynamics," *Circulation,* vol. 96, no. 3, pp. 842-8, Aug 5, 1997.
[23] P. Castiglioni, and M. Di Rienzo, "How the threshold "r" influences approximate entropy analysis of heart-rate variability." pp. 561-564.
[24] C. Y. Liu, C. C. Liu, P. Shao *et al.*, "Comparison of different threshold values r for approximate entropy: application to investigate the heart rate variability between heart failure and healthy control groups," *Physiological Measurement,* vol. 32, no. 2, pp. 167-180, Feb, 2011.
[25] P. Melillo, M. Bracale, and L. Pecchia, "Nonlinear Heart Rate Variability features for real-life stress detection. Case study: students under stress due to university examination," *Biomedical Engineering Online,* vol. 10, Nov 7, 2011.
[26] A. Jovic, and N. Bogunovic, "Electrocardiogram analysis using a combination of statistical, geometric, and nonlinear heart rate variability features," *Artificial Intelligence in Medicine,* vol. 51, no. 3, pp. 175-186, 2011.
[27] B. Singh, D. Singh, A. K. Jaryal *et al.*, "Ectopic beats in approximate entropy and sample entropy-based HRV assessment," *International Journal of Systems Science,* vol. 43, no. 5, pp. 884-893, 2012/05/01, 2011.
[28] V. Papaioannou, M. Giannakou, N. Maglaveras *et al.*, "Investigation of heart rate and blood pressure variability, baroreflex sensitivity, and approximate entropy in acute brain injury patients," *J Crit Care,* vol. 23, no. 3, pp. 380-6, Sep, 2008.
[29] B. B. Mandelbrot, *The Fractal Geometry of Nature*: Henry Holt and Company, 1982.
[30] T. Higuchi, "Approach to an Irregular Time Series on the Basis of the Fractal Theory," *Physica,* vol. 31, pp. 277-283, 1988, 1988.
[31] A. M. Fraser, and H. L. Swinney, "Independent coordinates for strange attractors from mutual information," *Physical Review A,* vol. 33, no. 2, pp. 1134-1140, 1986.
[32] A. M. Fraser, "Information and entropy in strange attractors," *Information Theory, IEEE Transactions on,* vol. 35, no. 2, pp. 245-262, 1989.
[33] P. Grassberger, and I. Procaccia, "Measuring the strangeness of strange attractors," *Physica D,* vol. 9, pp. 189 - 208, 1983.

[34] R. Acharya, A. Kumar, P. Bhat *et al.*, "Classification of cardiac abnormalities using heart rate signals," *Medical and Biological Engineering and Computing,* vol. 42, no. 3, pp. 288-293, 2004.

[35] R. Acharya, N. Kannathal, and S. M. Krishnan, "Comprehensive analysis of cardiac health using heart rate signals," *Physiological Measurement,* vol. 25, no. 5, pp. 1139-1151, Oct, 2004.

[36] H. V. Huikuri, T. H. Makikallio, C. K. Peng *et al.*, "Fractal correlation properties of R-R interval dynamics and mortality in patients with depressed left ventricular function after an acute myocardial infarction," *Circulation,* vol. 101, no. 1, pp. 47-53, Jan 4-11, 2000.

[37] S. Schulz, M. Koschke, K. J. Bar *et al.*, "The altered complexity of cardiovascular regulation in depressed patients," *Physiol Meas,* vol. 31, no. 3, pp. 303-21, Mar, 2010.

[38] C. K. Peng, S. Havlin, H. E. Stanley *et al.*, "Quantification of scaling exponents and crossover phenomena in nonstationary heartbeat time series," *Chaos: An Interdisciplinary Journal of Nonlinear Science,* vol. 5, no. 1, pp. 82-87, 1995.

[39] J. Eckmann, S. Kamphorst, and D. Ruelle, "Recurrence plots of dynamical systems," *Europhysics Letters,* vol. 4, pp. 973 - 977, 1987.

[40] H. Dabire, D. Mestivier, J. Jarnet *et al.*, "Quantification of sympathetic and parasympathetic tones by nonlinear indexes in normotensive rats," *American Journal of Physiology-Heart and Circulatory Physiology,* vol. 275, no. 4, pp. H1290-H1297, Oct, 1998.

[41] J. Theiler, S. Eubank, A. Longtin *et al.*, "Testing for nonlinearity in time series: the method of surrogate data," *Physica D,* vol. 58, pp. 77 - 94, 1992.

[42] N. Marwan, N. Wessel, U. Meyerfeldt, A. Schirdewan, J. Kurths, "Recurrence-plot-based measures of complexity and their application to heart-rate-variability data". *Phys Rev E Stat Nonlin Soft Matter Phys,* vol. 66, 2002.

[43] K. Natarajan, R. Acharya U, F. Alias *et al.*, "Nonlinear analysis of EEG signals at different mental states," *Biomedical Engineering Online,* vol. 3, no. 1, pp. 7, 2004.

In: Focus on Nonlinear Analysis Research
Editors: Gualtiero Padovani and Manuel Occhino
ISBN: 978-1-62417-350-9
© 2013 Nova Science Publishers, Inc.

Chapter 7

NONLINEAR EVOLUTION ANALYSIS FOR FLOW DURING ALLOY CRYSTALLIZATION

Dambaru Bhatta * ***and Daniel N. Riahi*** †
The University of Texas-Pan American, Edinburg, TX, US

Abstract

This article presents an analysis of nonlinear evolution equation appearing in fluid flow during solidification of a binary alloy which is cooled from below. It has been well established by experimentalists that a horizontal mushy layer is formed during the solidification of binary alloys. We have included a review section on recent studies done by various researchers on this topic. This chapter includes an analysis on nonlinear behavior of the convective flow for a passive as well as an active mushy layer. The effects of presence and absence of a magnetic field in the mushy layer are also considered. The basic state, linear, adjoint and first-order systems are derived and these solutions are used to obtain the Landau coefficients which appear in the resulting nonlinear evolution equation. Our computational results suggest that higher permeability parameter and Chandrasekhar number enhance the marginal stability. This analysis also indicates a possible slow transition to turbulence in such flow system since the computed Landau constants are positive.

Keywords: Nonlinear evolution, Convective flow, Mushy layer, Solidification, Landau constant, Magneto-convection, Permeability

AMS Subject Classification: 35Q80, 76E06, 76E30, 76S05

1. Introduction

Hydrodynamic stability problems were initially recognized and studied in the nineteenth century by various famous scientists including Helmholtz, Kelvin, Rayleigh and Reynolds. The first technique used to study hydrodynamic stability was linearization for small perturbation of basic flow. Weakly nonlinear study is based on linear theory which considers

*E-mail address: bhattad@utpa.edu
†E-mail address: driahi@utpa.edu

the leading nonlinear effects of small perturbations. Weakly nonlinear theory has been developed and used during in the twentieth century for various applied problems in hydrodynamic stability analysis. Lev Landau [1] proposed an equation to analyze hydrodynamic stability in 1944. This equation was named after him. The Landau equation has been derived for various cases [2] and the dependence of Landau constant for supercritical stability and subcritical stability has been discussed. Chandrasekhar [3] carried out extensive research work on hydrodynamic and hydromagnetic stability from 1950 to 1961. Area of solidification is becoming of great importance in real world applications with the advancement of technologies day by day. Solidification process is used in various fields such as manufacturing, material science, earth science, energy and environmental science. Crystal growth, ceramics, welding, casting are various applications in material science involving solidification. Volcanic systems, crystal magmas, ore deposits, sea ice are earth science related solidification applications. Experimentalists observed vertical chimneys or channels void of solid during solidification of binary alloys. A mushy zone or mushy layer, which is usually modeled as a porous medium, is partially solidified region that is formed during the solidification of alloys. Solidification of melts is greatly influenced by convection in mushy layer. Convective flows in mushy layer can produce chimney. It is also known that the convective flow within the chimneys can produce freckles in the final form of the solidified alloy. 'Freckles' are imperfections that reduce the quality of the solidified materials. It is important to investigate and discover the means of suppressing the freckle formation and hence produce single crystals of high quality for the semiconductor industry. Hills et al. [4] developed a set of thermodynamic equations for a mushy layer and solved the one dimensional freezing problem. Fowler [5] developed a mathematical model for the convective flow in chimneys and predicted a criterion for the onset of convection and freckling. Huppert [6] studied experimentally, the effects associated with cooling and solidifying a melt from below and also evaluated the criterion for which solid-liquid interface becomes unstable. Worster [7] considered a model for dendritic growth during the solidification of alloys, treated the mixed phase of solid and liquid as continuum and suggested different morphologies that can occur due to the variation of solid fraction. Vives et al. [8] carried out experimental studies of natural and damped convection during the solidification of tin and aluminum alloys caused by stationary and uniform magnetic field parallel to the gravity. They observed that the stationary magnetic field decreases the superheat and increases the rate of solidification. Many interesting and important, theoretical and experimental, investigations, concerning the convection within the mushy layer, which is responsible for the formation of chimneys, were reported during last three decades. Worster [9] developed mathematical model for the fluid flow in mushy layer and determine a critical Rayleigh number above which the mushy layer is unstable to small disturbances. Chen et al. [10] conducted directional solidification experiments using ammonium chloride-water solution, studied different types of convection and calculated the critical Rayleigh number of mushy layer for the onset of plume convection. The linear instability problem for a binary-alloy solidification system was solved by Worster [11]. He investigated two distinct modes of instability, one of which has been referred to as the mushy-layer mode. This mode is responsible for the development of chimneys within the mushy layer. These results were obtained under the restriction of infinite Prandtl number. The mushy-layer mode drives the convective flow downward everywhere except in and near the localized chimneys. This result is in agreement with the experimental result of Tait et al. [12], in which they conducted experiments using aqueous solutions of ammonium chloride. They observed that the boundary layer at the top of the mush become unstable, leading to the convection in the overlying fluid. They also observed that up flow occurred at the cells boundaries and the down flow at the cells centers. The simplified mushy-layer model was introduced first by Amberg et al. [13]. The model was based on a near eutectic approximation and in the limit of large far-field temperature. Such asymptotic limits allowed them to examine the dynamics of mushy layer in the form of a small deviation from the classical system of convection in horizontal porous medium with constant permeability. Emms et al. [14] proposed a coupled mush-liquid model to study the onset of convection during the directional solidification of binary alloys. Several attempts were made to study the solidification process by imposing the certain external constraints, such as rotation and/or magnetic fields, to suppress or at least reduce convection in mushy layers. Anderson et al. [15] employed a weakly nonlinear analysis of simplified mushy layer model that was proposed by Amberg et al. A near eutectic approximation was applied and the limit of large far-field temperature was considered. Such asymptotic limits allowed them to examine the dynamics of mushy layer. They also considered the limit of large Stefan number, which enabled them to reach a domain for the existence of the oscillatory mode of convection. Chen [16] carried out some experimental studies on convection in a mushy layer during directional solidification.

A number of examples can be found in the literature where external forces have been applied to weaken or to control the convection in solidifying mushy layers. One such example is to study the hydromagnetic effects on fluid flow in mushy regions, which has wide range of application in nuclear engineering and crystal growth. Riahi [17] investigated the role of nonlinear natural convection in cylindrical chimneys within a mushy layer during solidification of binary alloy under high gravity environment, where the rotation axis inclined with high gravity vector. It was observed that, for some moderate values of the rotation rate, axial convection in the chimneys decreases rapidly with increasing acceleration parameter and increasing Coriolis parameter. The effect of vertical magnetic field on the convection in solidifying mushy layer was studied asymptotically by Riahi [18]. Three cases of strong, moderate and weak magnetic fields were studied. The presence of the externally imposed magnetic field in the vertical direction was found to be stabilizing in the case of strong field and was found to be ineffective in the case of moderate (or weak) field in the leading order effects in the asymptotic limit of sufficiently large Rayleigh number.

Roper et al. [19] presented an analysis of convection in a mushy layer with deformable permeable interface. Okhuysen et al. [20, 21] examined a weakly nonlinear analysis of buoyant convection in binary alloy solidification for permeable mush-liquid interface and in the absence of vertical magnetic field. They analyzed effects of several parameters on two and three-dimensional steady convection patterns in the mushy layers for variable permeability. The most important result of their study was the prediction of subcritical down-hexagonal pattern for variable permeability case that corresponds to the smallest value of the Rayleigh number. Muddamallappa et al [22] used a modified mushy layer model based on the standard near eutectic approximation. They used linear stability analysis and calculated critical Rayleigh number for the cases of both constant and variable permeability. However the issues of nonlinear effects and transition effects on the chimney formation did not feature in their investigation. Study on oscillatory modes of nonlinear compositional convection in mushy layers was carried out by Riahi [23]. Guba et al. [24] studied interactions between steady and oscillatory convection. Bhatta et al. [25, 26, 27] derived Landau type equation using weakly nonlinear studies for a convective flow in mushy layer with and without magnetic field.

In section two, we present the governing equations with boundary conditions and a solution procedure for the mushy layer system. Derivations of the linear, first-order and second-order systems are presented in section three and section four deals with the derivation of Landau type equation for the amplitude. We discuss our numerical results in section five and conclusion in section six.

2. Mushy Layer with a Permeable Mush-Liquid Interface

We consider the domain of mushy layer (similar to Worster [9]) during solidification of binary alloy under externally imposed magnetic field. A mushy layer (figure 1), with an underlying solid and overlying liquid, is partially solidified region, known to produce convection. The solidification front is advancing at a speed of V_0 which is assumed to be constant. The liquid far above the mushy layer is assumed to have the temperature $T_\infty > T_L$,, where T_L is the liquidus temperature of the alloy. The mushy layer is assumed to be in the state of thermodynamic equilibrium ([9]), therefore $T = T_L(C_0) + \Gamma(C - C_0)$ where T is the temperature, C is the composition and Γ is the constant slope of the liquidus, C_0 is the composition of liquid.

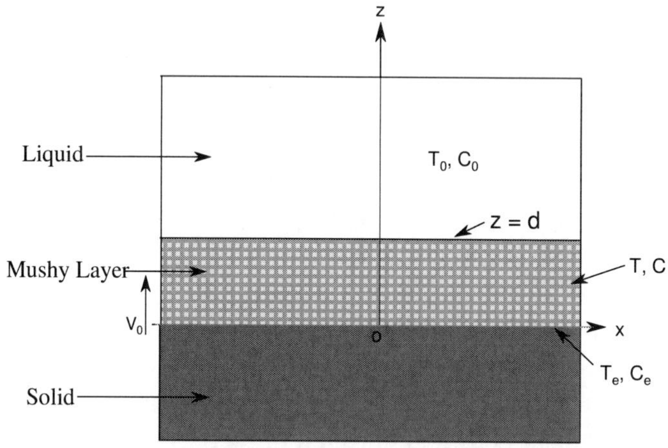

Figure 1. Schematic of the temperature and concentration fields and geometry for a mushy layer formed during solidification.

The nondimensional system can then be expressed as

$$\begin{aligned}
\mathcal{K}\overrightarrow{\mathcal{U}} + \nabla\mathcal{P} + \mathcal{R}\Theta\overrightarrow{k} &= 0 \\
\nabla\cdot\overrightarrow{\mathcal{U}} &= 0 \\
\left(\frac{\partial}{\partial t} - \frac{\partial}{\partial z}\right)[\Theta - \mathcal{S}\Phi] + \overrightarrow{\mathcal{U}}\cdot\nabla\Theta &= \nabla^2\Theta \\
\left(\frac{\partial}{\partial t} - \frac{\partial}{\partial z}\right)[(1-\Phi)\Theta + \mathcal{C}\Phi] + \overrightarrow{\mathcal{U}}\cdot\nabla\Theta &= 0
\end{aligned} \tag{1}$$

with boundary conditions:

$$\begin{aligned}
\Theta = -1,\ \mathcal{W} = 0 \quad &\text{at} \quad z = 0 \\
\Theta = \Phi = \partial_z\mathcal{W} = 0 \quad &\text{at} \quad z = \delta
\end{aligned}$$

where $\delta = dV_0/\kappa$ represents the dimensionless depth of the mushy layer and $\mathcal{W}$ denotes the vertical component of $\overrightarrow{\mathcal{U}}$.

Here Θ, Φ, $\overrightarrow{\mathcal{U}}$, $\mathcal{P}$ and $\overrightarrow{\mathcal{H}}$ are the non dimensional dependent variables representing heat, solid fraction, liquid flux and pressure. Independent variables are space variables x, y, z and time variable t. Non dimensional numbers appearing in the model are Stefan number $\mathcal{S}$, concentration ratio $\mathcal{C}$ and Rayleigh number $\mathcal{R}$.

2.1. Solution Procedure

Assuming solutions of the form

$$\begin{aligned}
\Theta(x,y,z,t) &= \theta_b(z) + \theta\,(x,y,z,t) \\
\Phi(x,y,z,t) &= \phi_b(z) + \phi\,(x,y,z,t) \\
\overrightarrow{\mathcal{U}}(x,y,z,t) &= \overrightarrow{0} + \overrightarrow{u}\,(x,y,z,t) \\
\mathcal{P}(x,y,z,t) &= p_b(z) + p\,(x,y,z,t) \\
\mathcal{K}(x,y,z,t) &= k_b(z) + K\,(x,y,z,t)
\end{aligned} \tag{2}$$

where θ_b, ϕ_b, p_b, k_b are dependent variables for the steady basic state motionless system, θ, ϕ, $\overrightarrow{u}$, p, K are dependent variables for the perturbation which are superimposed on the basic state system.

2.2. Basic State Solutions

Using (2) in (1) and in the absence of perturbation, we obtain steady basic state system as

$$\frac{d^2\theta_b}{dz^2} + \frac{d\theta_b}{dz} - \mathcal{S}\frac{d\phi_b}{dz} = 0 \tag{3}$$

$$(1-\phi_b)\frac{d\theta_b}{dz} + (\mathcal{C} - \theta_b)\frac{d\phi_b}{dz} = 0 \tag{4}$$

$$\frac{dp_b}{dz} + \mathcal{R}\theta_b = 0 \tag{5}$$

with boundary conditions:

$$\theta_b = -1 \qquad \text{at} \quad z = 0$$
$$\theta_b = \phi_b = 0 \qquad \text{at} \quad z = \delta.$$

Solution θ_b in implicit form is given by

$$z = \frac{\alpha_1 - \mathcal{C}}{\alpha_1 - \alpha_2} \ln\left[\frac{1+\alpha_1}{\alpha_1 - \theta_b}\right] + \frac{\mathcal{C} - \alpha_2}{\alpha_1 - \alpha_2} \ln\left[\frac{1+\alpha_2}{\alpha_2 - \theta_b}\right] \tag{6}$$

Here, α_1 and α_2 are, respectively, given by

$$\alpha_1 = \frac{\mathcal{C} + \mathcal{S} + \theta_\infty + \sqrt{(\mathcal{C} + \mathcal{S} + \theta_\infty)^2 - 4\mathcal{C}\theta_\infty}}{2}$$
$$\alpha_2 = \frac{\mathcal{C} + \mathcal{S} + \theta_\infty - \sqrt{(\mathcal{C} + \mathcal{S} + \theta_\infty)^2 - 4\mathcal{C}\theta_\infty}}{2}$$

where θ_∞ is the non-dimensional temperature far away from mush/liquid interface and solution ϕ_b is given by

$$\phi_b = \frac{\theta_b}{\theta_b - \mathcal{C}} \tag{7}$$

and k_b depends on ϕ_b. Thickness of the layer can be determined as

$$\delta = \frac{\alpha_1 - \mathcal{C}}{\alpha_1 - \alpha_2} \ln\left[\frac{1+\alpha_1}{\alpha_1}\right] + \frac{\mathcal{C} - \alpha_2}{\alpha_1 - \alpha_2} \ln\left[\frac{1+\alpha_2}{\alpha_2}\right].$$

2.3. Perturbed System

Following equations are obtained for the perturbation quantities:

$$\begin{aligned}
&k_b\vec{u} + \nabla p + \mathcal{R}\theta\vec{k} = -K\vec{u} \\
&\nabla\cdot\vec{u} = 0 \\
&\left(\frac{\partial}{\partial t} - \frac{\partial}{\partial z} - \nabla^2\right)\theta - \mathcal{S}\left(\frac{\partial}{\partial t} - \frac{\partial}{\partial z}\right)\phi + w\frac{d\theta_b}{dz} = -\left(\vec{u}\cdot\nabla\theta\right) \\
&\left(\frac{\partial}{\partial t} - \frac{\partial}{\partial z}\right)\{(1-\phi_b)\theta + (\mathcal{C} - \theta_b)\phi\} + w\frac{d\theta_b}{dz} \\
&\qquad = \left\{-\vec{u}\cdot\nabla\theta + \left(\frac{\partial}{\partial t} - \frac{\partial}{\partial z}\right)(\theta\phi)\right\}
\end{aligned} \tag{8}$$

with boundary conditions

$$\theta = w = 0 \qquad \text{at} \quad z = 0$$
$$\theta = \phi = \partial_z w = 0 \qquad \text{at} \quad z = \delta.$$

Writing $\overrightarrow{u} = (u, v, w) = \nabla \times \nabla \times (u_P\vec{k}) + \nabla \times (u_T\vec{k})$, where u_P and u_T represent poloidal and toroidal components of $\overrightarrow{u}$. Here $\vec{k}$ is the unit vector in the positive z-direction. Ignoring u_T [3] as it has no effect on our two-dimensional case, we obtain

$$(u, v, w) = \left(\frac{\partial^2 u_P}{\partial x \partial z}, \frac{\partial^2 u_P}{\partial y \partial z}, -\Delta_2 u_P\right).$$

Here Δ_2 is used to denote the Laplacian in the xy-plane. Now we eliminate the pressure term from the first equation of the system (8) by taking the double curl. Thus, using the continuity equation, i.e., the second equation of the system (8), the third component of $\nabla \times \nabla \times (k_b \overrightarrow{u})$ can be written as

$$\begin{aligned}
&\frac{\partial^2(k_b u)}{\partial x \partial z} + \frac{\partial^2(k_b v)}{\partial y \partial z} - \frac{\partial^2(k_b w)}{\partial x^2} - \frac{\partial^2(k_b w)}{\partial y^2} \\
&= -\frac{\partial k_b}{\partial z}\frac{\partial w}{\partial z} - k_b\frac{\partial^2 w}{\partial z^2} - k_b\frac{\partial^2 w}{\partial x^2} - k_b\frac{\partial^2 w}{\partial y^2} \\
&= -\frac{\partial k_b}{\partial z}\frac{\partial w}{\partial z} - k_b\nabla^2 w = \frac{\partial k_b}{\partial z}\frac{\partial}{\partial z}(\Delta_2 u_P) + k_b\nabla^2(\Delta_2 u_P).
\end{aligned}$$

Similarly, for $\nabla \times \nabla \times (\mathcal{R}\theta \overrightarrow{k})$, we have

$$-\mathcal{R}\left[\frac{\partial^2\theta}{\partial x^2} + \frac{\partial^2\theta}{\partial y^2}\right] = -\mathcal{R}(\Delta_2\theta)$$

and for $\nabla \times \nabla \times (K\overrightarrow{u})$, we have

$$\begin{aligned}
&\frac{\partial^2(Ku)}{\partial x \partial z} + \frac{\partial^2(Kv)}{\partial y \partial z} - \frac{\partial^2(Kw)}{\partial x^2} - \frac{\partial^2(Kw)}{\partial y^2} \\
&= \frac{\partial^2}{\partial x \partial z}\left(K\frac{\partial^2 u_P}{\partial x \partial z}\right) + \frac{\partial^2}{\partial y \partial z}\left(K\frac{\partial^2 u_P}{\partial y \partial z}\right) + \frac{\partial^2}{\partial x^2}(K\Delta_2 u_P) + \frac{\partial^2}{\partial y^2}(K\Delta_2 u_P).
\end{aligned}$$

Hence, the perturbed system in 2-D becomes

$$\begin{aligned}
&k_b\nabla^2(\Delta_2 u_P) - \mathcal{R}(\Delta_2\theta) + \frac{\partial k_b}{\partial z}\frac{\partial}{\partial z}(\Delta_2 u_P) \\
&= -\left[\frac{\partial^2}{\partial x^2}\{K(\Delta_2 u_P)\} + \frac{\partial^2}{\partial x \partial z}\left\{K\frac{\partial^2 u_P}{\partial x \partial z}\right\}\right] \qquad (9)
\end{aligned}$$

$$\begin{aligned}
&\left(\nabla^2 + \frac{\partial}{\partial z} - \frac{\partial}{\partial t}\right)\theta - \mathcal{S}\left(\frac{\partial}{\partial z} - \frac{\partial}{\partial t}\right)\phi + (\Delta_2 u_P)\frac{\partial\theta_b}{\partial z} \\
&= \left[\frac{\partial^2 u_P}{\partial x \partial z}\frac{\partial\theta}{\partial x} - (\Delta_2 u_P)\frac{\partial\theta}{\partial z}\right] \qquad (10)
\end{aligned}$$

$$\begin{aligned}
&\left(\frac{\partial}{\partial z} - \frac{\partial}{\partial t}\right)[(\theta_b - \mathcal{C})\phi - (1 - \phi_b)\theta] - (\Delta_2 u_P)\frac{\partial\theta_b}{\partial z} \\
&= -\left[\left(\frac{\partial}{\partial z} - \frac{\partial}{\partial t}\right)[\theta\phi] + \frac{\partial^2 u_P}{\partial x \partial z}\frac{\partial\theta}{\partial x} - (\Delta_2 u_P)\frac{\partial\theta}{\partial z}\right]. \qquad (11)
\end{aligned}$$

2.4. Derivation of Linear, First-order and Second-order Systems

Now we consider the linear, first-order and second-order solutions in 2-D for the perturbation whose amplitude designated by ϵ is assumed to be small ($\epsilon \ll 1$), by expressing as follows

$$
\begin{aligned}
u_P(x,z,t) &= \epsilon u_{P0}(x,z,t) + \epsilon^2 u_{P1}(x,z,t) + \epsilon^3 u_{P2}(x,z,t) + ...\\
\theta(x,z,t) &= \epsilon\theta_0(x,z,t) + \epsilon^2\theta_1(x,z,t) + \epsilon^3\theta_2(x,z,t) + ...\\
\phi(x,z,t) &= \epsilon\phi_0(x,z,t) + \epsilon^2\phi_1(x,z,t) + \epsilon^3\phi_2(x,z,t) + ...\\
K(x,z,t) &= \epsilon K_0(x,z,t) + \epsilon^2 K_1(x,z,t) + \epsilon^3 K_2(x,z,t) + ...
\end{aligned} \tag{12}
$$

where subscripts 0, 1 and 2 are used for linear, first-order and second-order perturbed quantities respectively. Following Drazin and Reid [2], we define slow time variable t_s given by

$$
t_s = \epsilon^2 t \qquad \text{with} \qquad \epsilon^2 = \frac{\mathcal{R} - R_0}{R_2} > 0. \tag{13}
$$

Here the sign and magnitude of R_2 is chosen so that ϵ^2 remains small and positive. R_0 is the value of the Rayleigh number at the onset of motion for the linear system. The expansions are used to study the system at small perturbation above the onset of instability.

2.4.1. Linear System

The linear system can be obtained by substituting (13) into (9), (10) and (11) and collecting coefficients of the $O(\epsilon)$ terms as

$$
\begin{aligned}
&k_b\nabla^2\left(\Delta_2 u_{P0}\right) - R_0\left(\Delta_2\theta_0\right) + \left(\partial_z k_b\right)\partial_z\left(\Delta_2 u_{P0}\right) = 0\\
&\left(\nabla^2 + \partial_z - \partial_t\right)\theta_0 - \mathcal{S}\left(\partial_z - \partial_t\right)\phi_0 + \left(\partial_z\theta_b\right)\left(\Delta_2 u_{P0}\right) = 0\\
&\left(\partial_z - \partial_t\right)\left[\left(\theta_b - \mathcal{C}\right)\phi_0 - \left(1 - \phi_b\right)\theta_0\right] - \left(\partial_z\theta_b\right)\left(\Delta_2 u_{P0}\right) = 0
\end{aligned} \tag{14}
$$

with boundary conditions:

$$
\begin{aligned}
&u_{P0} = \theta_0 = 0 \qquad \text{at} \quad z = 0\\
&\partial_z u_{P0} = \theta_0 = \phi_0 = 0 \qquad \text{at} \quad z = \delta.
\end{aligned}
$$

We assume that solutions can be written in terms of normal modes [2], i.e.,

$$
\begin{aligned}
u_{P0}\left(x,z,t_s\right) &= A(t_s)\tilde{P_0}(z)e^{i\alpha x} + \text{C.C.}\\
\theta_0\left(x,z,t_s\right) &= A(t_s)\tilde{\theta_0}(z)e^{i\alpha x} + \text{C.C.}\\
\phi_0\left(x,z,t_s\right) &= A(t_s)\tilde{\phi_0}(z)e^{i\alpha x} + \text{C.C.}\\
K_0\left(x,z,t_s\right) &= A(t_s)\tilde{K_0}(z)e^{i\alpha x} + \text{C.C.}
\end{aligned}
$$

where C.C. stands for complex conjugate and (14) becomes

$$k_b\left[D^2-\alpha^2\right]\tilde{P_0}-R_0\tilde{\theta_0}+(Dk_b)\left(D\tilde{P_0}\right)=0 \tag{15}$$

$$\left[D^2+D-\alpha^2\right]\tilde{\theta_0}-\mathcal{S}\left(D\tilde{\phi_0}\right)-(D\theta_b)\,\alpha^2\tilde{P_0}=0 \tag{16}$$

$$D\left[(\theta_b-\mathcal{C})\,\tilde{\phi_0}-(1-\phi_b)\,\tilde{\theta_0}\right]+(D\theta_b)\,\alpha^2\tilde{P_0}=0 \tag{17}$$

where $D=\partial_z$ which can be written as

$$Lq \;=\; 0 \tag{18}$$

with

$$L=\begin{bmatrix} k_bD^2+(Dk_b)\,D-\alpha^2k_b & -R_0 & 0 \\ -\alpha^2\,(D\theta_b) & D^2+D-\alpha^2 & -\mathcal{S}D \\ \alpha^2\,(D\theta_b) & -(1-\phi_b)\,D+(D\phi_b) & (\theta_b-\mathcal{C})\,D+(D\theta_b) \end{bmatrix} \tag{19}$$

and $q=\left[\begin{array}{ccc}\tilde{P_0} & \tilde{\theta_0} & \tilde{\phi_0}\end{array}\right]^T$. Here T is used to denote the transpose.

2.4.2. Adjoint System

Now we define the linear adjoint operator L^* of the linear operator L as

$$\langle Lq,\,\hat{q}\rangle \;=\; \langle q,\,L^*\hat{q}\rangle \tag{20}$$

where L is given by (19), $\hat{q}=\left[\begin{array}{ccc}\hat{u}_P & \hat{\theta} & \hat{\phi}\end{array}\right]^T$ and $\langle f,\,g\rangle \;=\; \int_0^\delta fg dz$.

To obtain the adjoint system, we multiply the equations (15), (16), and (17) by $\hat{u}_P,\ \hat{\theta}$ and $\hat{\phi}$ respectively, and then integrate with respect to z from $z=0$ to $z=\delta$. These yield

$$\begin{aligned}&\int_0^\delta \hat{u}_P\left\{k_b\left(D^2-\alpha^2\right)\tilde{P_0}-R_0\tilde{\theta_0}+(Dk_b)\,D\tilde{P_0}\right\}dz\\ &+\int_0^\delta \hat{\theta}\left\{\left(D^2+D-\alpha^2\right)\tilde{\theta_0}-\mathcal{S}\left(D\tilde{\phi_0}\right)-(D\theta_b)\,\alpha^2\tilde{P_0}\right\}dz\\ &+\int_0^\delta \hat{\phi}\left\{D\left[(\theta_b-\mathcal{C})\,\tilde{\phi_0}-(1-\phi_b)\,\tilde{\theta_0}\right]+(D\theta_b)\,\alpha^2\tilde{P_0}\right\}dz=0\end{aligned} \tag{21}$$

Boundary conditions for the adjoint system are

$$\begin{aligned}\hat{u}_P=\hat{\theta}=\hat{\phi}=0 \qquad &\text{at} \quad z=0\\ D\hat{u}_P=\hat{\theta}=0 \qquad &\text{at} \quad z=\delta.\end{aligned}$$

Integration by parts on (21), and use of the boundary conditions yield

$$\int_0^\delta \left\{D^2\left(\hat{u}_P k_b\right) - D\left(\hat{u}_P D k_b\right) - \alpha^2 \hat{u}_P k_b + \alpha^2\left(\hat{\phi} - \hat{\theta}\right)\left(D\theta_b\right)\right\} \tilde{P}_0 dz$$
$$+ \int_0^\delta \left\{D^2\hat{\theta} - D\hat{\theta} - R_0\hat{u}_P - \alpha^2\hat{\theta} + \left(1-\phi_b\right)\left(D\hat{\phi}\right)\right\}\tilde{\theta}_0 dz$$
$$+ \int_0^\delta \left\{\mathcal{S}\left(D\hat{\theta}\right) + \left(\mathcal{C} - \theta_b\right)\left(D\hat{\phi}\right)\right\}\tilde{\phi}_0 dz = 0$$

Thus the adjoint system can be expressed as

$$\begin{aligned} &\left\{k_b D^2 + \left(Dk_b\right)D - \alpha^2 k_b\right\}\hat{u}_P + \alpha^2\left(\hat{\phi} - \hat{\theta}\right)\left(D\theta_b\right) = 0 \\ &\left\{D^2 - D - \alpha^2\right\}\hat{\theta} - R_0\hat{u}_P + \left(1-\phi_b\right)\left(D\hat{\phi}\right) = 0 \\ &\mathcal{S}\left(D\hat{\theta}\right) + \left(\mathcal{C} - \theta_b\right)\left(D\hat{\phi}\right) = 0 \end{aligned} \tag{22}$$

and, in turn, we can write adjoint operator L^* as

$$L^* = \begin{bmatrix} k_b D^2 + \left(Dk_b\right)D - \alpha^2 k_b & -\alpha^2 D\theta_b & \alpha^2 D\theta_b \\ -R_0 & D^2 - D - \alpha^2 & \left(1-\phi_b\right)D \\ 0 & \mathcal{S}D & \left(\mathcal{C} - \theta_b\right)D \end{bmatrix} \tag{23}$$

2.4.3. First Order System

The first-order system is obtained by substituting (13) into (9), (10) and (11) and collecting coefficients of the $O(\epsilon)$ terms. The equations for the first-order problem become

$$\begin{aligned} &k_b\nabla^2\left(\Delta_2 u_{P1}\right) - R_0\left(\Delta_2\theta_1\right) + \frac{\partial k_b}{\partial z}\frac{\partial}{\partial z}\left\{\Delta_2 u_{P1}\right\} \\ &\quad = -\left[\frac{\partial^2}{\partial x^2}\left\{K_0\left(\Delta_2 u_{P0}\right)\right\} + \frac{\partial^2}{\partial x\partial z}\left\{K_0\frac{\partial^2 u_{P0}}{\partial x\partial z}\right\}\right] \end{aligned} \tag{24}$$

$$\begin{aligned} &\left(\nabla^2 + \frac{\partial}{\partial z}\right)\theta_1 - \mathcal{S}\frac{\partial\phi_1}{\partial z} + \frac{\partial\theta_b}{\partial z}\left(\Delta_2 u_{P1}\right) \\ &\quad = \frac{\partial^2 u_{P0}}{\partial x\partial z}\frac{\partial\theta_0}{\partial x} - \left(\Delta_2 u_{P0}\right)\frac{\partial\theta_0}{\partial z} \end{aligned} \tag{25}$$

$$\begin{aligned} &\frac{\partial}{\partial z}\left[\left(1-\phi_b\right)\theta_1 + \left(\mathcal{C} - \theta_b\right)\phi_1\right] + \frac{\partial\theta_b}{\partial z}\left(\Delta_2 u_{P1}\right) \\ &\quad = \frac{\partial}{\partial z}\left[\theta_0\phi_0\right] + \frac{\partial^2 u_{P0}}{\partial x\partial z}\frac{\partial\theta_0}{\partial x} - \left(\Delta_2 u_{P0}\right)\frac{\partial\theta_0}{\partial z} \end{aligned} \tag{26}$$

Using the normal mode, we write the first-order quantities as

$$\begin{aligned} u_{P1}(x,z,t_s) &= A^2(t_s)P_{12}(z)e^{2i\alpha x} + |A(t_s)|^2 P_{10}(z) + \text{C.C.} \\ \theta_1(x,z,t_s) &= A^2(t_s)\theta_{12}(z)e^{2i\alpha x} + |A(t_s)|^2\theta_{10}(z) + \text{C.C.} \\ \phi_1(x,z,t_s) &= A^2(t_s)\phi_{12}(z)e^{2i\alpha x} + |A(t_s)|^2\phi_{10}(z) + \text{C.C.} \\ K_1(x,z,t_s) &= A^2(t_s)K_{12}(z)e^{2i\alpha x} + |A(t_s)|^2 K_{10}(z) + \text{C.C.} \end{aligned}$$

where subscript 12 is used for first-order two$-\alpha$ mode and subscript 10 is used for first-order zero$-\alpha$ mode.

Two- α Mode

Two-α mode equations corresponding to equations (24), (25) and (26) become

$$\begin{aligned} &k_b\left(D^2-4\alpha^2\right)P_{12}-R_0\theta_{12}+(Dk_b)\,DP_{12}\\ &\quad=\alpha^2\tilde{K}_0\tilde{P}_0-\frac{1}{2}D\left(\tilde{K}_0D\tilde{P}_0\right) \qquad (27)\\ &\left(D^2+D-4\alpha^2\right)\theta_{12}-\mathcal{S}\left(D\phi_{12}\right)-4\alpha^2\left(D\theta_b\right)P_{12}\\ &\quad=\alpha^2\left(\tilde{P}_0D\tilde{\theta}_0-\tilde{\theta}_0D\tilde{P}_0\right) \qquad (28)\\ &\left[\left(1-\phi_b\right)D-\left(D\phi_b\right)\right]\theta_{12}+\left[\left(\mathcal{C}-\theta_b\right)D-\left(D\theta_b\right)\right]\phi_{12}-4\alpha^2\left(D\theta_b\right)P_{12}\\ &\quad=D\left(\tilde{\theta}_0\tilde{\phi}_0\right)+\alpha^2\left(\tilde{P}_0D\tilde{\theta}_0-\tilde{\theta}_0D\tilde{P}_0\right) \qquad (29)\end{aligned}$$

Zero- α Mode

Zero-α mode equations corresponding to equations (24), (25) and (26) become

$$\begin{aligned} &\Delta_2P_{10}=0 \qquad (30)\\ &\left(D^2+D\right)\theta_{10}-\mathcal{S}\left(D\phi_{10}\right)=\alpha^2D\left(\tilde{\theta}_0^*\tilde{P}_0+\tilde{\theta}_0\tilde{P}_0^*\right) \qquad (31)\\ &\left[\left(1-\phi_b\right)D-\left(D\phi_b\right)\right]\theta_{10}+\left[\left(\mathcal{C}-\theta_b\right)D-\left(D\theta_b\right)\right]\phi_{10}\\ &\quad=D\left(\tilde{\theta}_0^*\tilde{\phi}_0+\tilde{\theta}_0\tilde{\phi}_0^*\right)+\alpha^2D\left(\tilde{\theta}_0^*\tilde{P}_0+\tilde{\theta}_0\tilde{P}_0^*\right) \qquad (32)\end{aligned}$$

2.4.4. Second Order System

For the second-order problem, the equation (9) becomes

$$\begin{aligned} &k_b\nabla^2\left(\Delta_2u_{P2}\right)-R_0\left(\Delta_2\theta_2\right)-R_2\left(\Delta_2\theta_0\right)+\frac{\partial k_b}{\partial z}\frac{\partial}{\partial z}\left(\Delta_2u_{P2}\right)=-\left[\frac{\partial^2}{\partial x^2}\left\{K_0\left(\Delta_2u_{P1}\right)\right\}\right.\\ &\left.+\frac{\partial^2}{\partial x^2}\left\{K_1\left(\Delta_2u_{P0}\right)\right\}+\frac{\partial^2}{\partial x\partial z}\left\{K_0\frac{\partial^2u_{P1}}{\partial x\partial z}\right\}+\frac{\partial^2}{\partial x\partial z}\left\{K_1\frac{\partial^2u_{P0}}{\partial x\partial z}\right\}\right] \qquad (33)\end{aligned}$$

Equation (10) transforms to

$$\begin{aligned} &\frac{\partial\theta_0}{\partial t_s}-\left(\nabla^2+\frac{\partial}{\partial z}\right)\theta_2-\mathcal{S}\left(\frac{\partial\phi_0}{\partial t_s}-\frac{\partial\phi_2}{\partial z}\right)-\left(\Delta_2u_{P2}\right)\frac{\partial\theta_b}{\partial z}\\ &=\left(\Delta_2u_{P0}\right)\frac{\partial\theta_1}{\partial z}+\left(\Delta_2u_{P1}\right)\frac{\partial\theta_0}{\partial z}-\frac{\partial^2u_{P0}}{\partial x\partial z}\frac{\partial\theta_1}{\partial x}-\frac{\partial^2u_{P1}}{\partial x\partial z}\frac{\partial\theta_0}{\partial x} \qquad (34)\end{aligned}$$

Equation (11) yields

$$(1-\phi_b)\frac{\partial\theta_0}{\partial t_s}-\frac{\partial}{\partial z}\left[(1-\phi_b)\,\theta_2\right]+(\mathcal{C}-\theta_b)\frac{\partial\phi_0}{\partial t_s}-\frac{\partial}{\partial z}\left[(\mathcal{C}-\theta_b)\,\phi_2\right]-(\Delta_2 u_{P2})\frac{\partial\theta_b}{\partial z}$$
$$=(\Delta_2 u_{P0})\frac{\partial\theta_1}{\partial z}+(\Delta_2 u_{P1})\frac{\partial\theta_0}{\partial z}-\frac{\partial}{\partial z}(\theta_0\phi_1+\theta_1\phi_0)-\frac{\partial^2 u_{P0}}{\partial x\partial z}\frac{\partial\theta_1}{\partial x}-\frac{\partial^2 u_{P1}}{\partial x\partial z}\frac{\partial\theta_0}{\partial x} \tag{35}$$

Using the normal mode, we have

$$\begin{aligned}
u_{P2}\,(x,z,t_s) &= \tilde{P_2}(z,\,t_s)e^{i\alpha x}+\text{C.C.}\\
\theta_2\,(x,z,t_s) &= \tilde{\theta_2}(z,\,t_s)e^{i\alpha x}+\text{C.C.}\\
\phi_2\,(x,z,t_s) &= \tilde{\phi_2}(z,\,t_s)e^{i\alpha x}+\text{C.C.}\\
K_2\,(x,z,t_s) &= \tilde{K_2}(z,\,t_s)e^{i\alpha x}+\text{C.C.}
\end{aligned}$$

Then the above three equations become

$$\begin{aligned}
&k_b\left(D^2-\alpha^2\right)\tilde{P_2}-R_0\tilde{\theta_2}-R_2\tilde{\theta_0}A+(Dk_b)\left(D\tilde{P_2}\right)\\
&=A|A|^2\left\{\alpha^2\left(4\tilde{K_0^*}P_{12}+\tilde{P_0^*}K_{12}+\tilde{P_0}(K_{10}+K_{10}^*)\right)\right.\\
&\left.+D\left(K_{12}(D\tilde{P_0^*})-2\tilde{K_0^*}(DP_{12})-(K_{10}+K_{10}^*)(D\tilde{P_0})\right)\right\}
\end{aligned} \tag{36}$$

and

$$\begin{aligned}
&\left(D^2+D-\alpha^2\right)\tilde{\theta_2}+\frac{dA}{dt_s}\left(\mathcal{S}\tilde{\phi_0}-\tilde{\theta_0}\right)-\mathcal{S}\left(D\tilde{\phi_2}\right)-\alpha^2\left(D\theta_b\right)\tilde{P_2}=A|A|^2\alpha^2\\
&\times\left\{2\left(\theta_{12}(D\tilde{P_0^*})+\tilde{\theta_0^*}(DP_{12})+2P_{12}(D\tilde{\theta_0^*})\right)+\tilde{P_0^*}(D\theta_{12})+\tilde{P_0}D(\theta_{10}+\theta_{10}^*)\right\}
\end{aligned} \tag{37}$$

and

$$\begin{aligned}
&(1-\phi_b)\left[\frac{dA}{dt_s}\tilde{\theta_0}-D\tilde{\theta_2}\right]+(\mathcal{C}-\theta_b)\left[\frac{dA}{dt_s}\tilde{\phi_0}-D\tilde{\phi_2}\right]+(D\phi_b)\,\tilde{\theta_2}+(D\theta_b)\left[\tilde{\phi_2}+\alpha^2\tilde{P_2}\right]\\
&=-A|A|^2\left[D\left(\tilde{\theta_0^*}\phi_{12}+\tilde{\phi_0^*}\tilde{\theta_{12}}+\tilde{\theta_0}(\phi_{10}+\phi_{10}^*)+\tilde{\phi_0}(\theta_{10}+\theta_{10}^*)\right)+\alpha^2\left\{2\left(\theta_{12}(D\tilde{P_0^*})\right.\right.\right.\\
&\left.\left.\left.+\tilde{\theta_0^*}(DP_{12})+2P_{12}(D\tilde{\theta_0^*})\right)+\tilde{P_0^*}(D\theta_{12})+\tilde{P_0}D(\theta_{10}+\theta_{10}^*)\right\}\right]
\end{aligned} \tag{38}$$

2.5. Derivation of the Landau-type Equation

Now, we multiply equation (36) by P_a, equation (37) by $\hat{\theta}$, and equation (38) by $\hat{\phi}$, and then integrate with respect to z from $z=0$ to $z=\delta$ to obtain an equation whose left hand side (LHS) and right hand side (RHS) are given by

$$
\begin{aligned}
LHS &= \int_0^\delta \hat{u}_P k_b \left(D^2-\alpha^2\right)\tilde{P}_2 dz - R_0\int_0^\delta \hat{u}_P\tilde{\theta}_2 dz - R_2 A\int_0^\delta \hat{u}_P\tilde{\theta}_0 dz \\
&+\int_0^\delta \hat{u}_P\left(Dk_b\right)\left(D\tilde{P}_2\right)dz + \int_0^\delta \hat{\theta}\left(D^2+D-\alpha^2\right)\tilde{\theta}_2 dz + \frac{dA}{dt_s}\int_0^\delta \hat{\theta}\left(\mathcal{S}\tilde{\phi}_0-\tilde{\theta}_0\right)dz \\
&-\mathcal{S}\int_0^\delta \hat{\theta}\left(D\tilde{\phi}_2\right)dz - \alpha^2\int_0^\delta \hat{\theta}\left(D\theta_b\right)\tilde{P}_2 dz + \frac{dA}{dt_s}\int_0^\delta \hat{\phi}\left(1-\phi_b\right)\tilde{\theta}_0 dz \\
&-\int_0^\delta \hat{\phi}\left(1-\phi_b\right)D\tilde{\theta}_2 dz + \frac{dA}{dt_s}\int_0^\delta \hat{\phi}\left(\mathcal{C}-\theta_b\right)\tilde{\phi}_0 dz - \int_0^\delta \hat{\phi}\left(\mathcal{C}-\theta_b\right)D\tilde{\phi}_2 dz \\
&+\int_0^\delta \hat{\phi}\left(D\phi_b\right)\tilde{\theta}_2 dz + \int_0^\delta \hat{\phi}\left(D\theta_b\right)\left[\tilde{\phi}_2+\alpha^2\tilde{P}_2\right]dz + \alpha^2\int_0^\delta \hat{\phi}\left(D\theta_b\right)\tilde{P}_2 dz
\end{aligned}
$$

$$
\begin{aligned}
RHS &= A|A|^2\int_0^\delta \Big[\hat{u}_P\Big\{D\Big(K_{12}(D\tilde{P}_0^*) - 2\tilde{K}_0^*(DP_{12}) - (K_{10}+K_{10}^*)(D\tilde{P}_0)\Big) \\
&+\alpha^2\Big(4\tilde{K}_0^* P_{12} + \tilde{P}_0^* K_{12} + \tilde{P}_0(K_{10}+K_{10}^*)\Big)\Big\} + \alpha^2\hat{\theta}\Big\{2\theta_{12}(D\tilde{P}_0^*) + 2\tilde{\theta}_0^*(DP_{12}) \\
&+4P_{12}(D\tilde{\theta}_0^*) + \tilde{P}_0 D(\theta_{10}+\theta_{10}^*) + \tilde{P}_0^*(D\theta_{12})\Big\} - \hat{\phi}\Big\{D\Big(\tilde{\theta}_0^*\phi_{12} + \tilde{\theta}_0(\phi_{10}+\phi_{10}^*) \\
&+\tilde{\phi}_0(\theta_{10}+\theta_{10}^*) + \tilde{\phi}_0^*\theta_{12}\Big) + \alpha^2\Big(\tilde{P}_0^*(D\theta_{12}) + 2\theta_{12}(D\tilde{P}_0^*) \\
&+2\tilde{\theta}_0^*(DP_{12}) + 4P_{12}(D\tilde{\theta}_0^*) + \tilde{P}_0 D(\theta_{10}+\theta_{10}^*)\Big)\Big\}\Big]dz
\end{aligned}
$$

Use of integration by parts and the boundary conditions, LHS yields

$$
\frac{dA}{dt_s}\int_0^\delta \left[\hat{\theta}\left(\mathcal{S}\tilde{\phi}_0-\tilde{\theta}_0\right) + \hat{\phi}\left(1-\phi_b\right)\tilde{\theta}_0 + \hat{\phi}\left(\mathcal{C}-\theta_b\right)\tilde{\phi}_0\right]dz - R_2 A\int_0^\delta \hat{u}_P\tilde{\theta}_0 dz
$$

Thus, equating LHS and RHS, we have

$$
T_1\frac{dA}{dt_s} = T_2 A + T_3 A|A|^2 \tag{39}
$$

where the terms T_1, T_2 and T_3 are given by

$$
T_1 = \int_0^\delta \left[\hat{\theta}\left(\mathcal{S}\tilde{\phi}_0-\tilde{\theta}_0\right) + \hat{\phi}\left(1-\phi_b\right)\tilde{\theta}_0 + \hat{\phi}\left(\mathcal{C}-\theta_b\right)\tilde{\phi}_0\right]dz \tag{40}
$$

$$
T_2 = R_2\int_0^\delta \hat{u}_P\tilde{\theta}_0 dz \tag{41}
$$

$$\begin{aligned} T_3 &= \int_0^\delta \Big[\Big(D \Big\{ K_{12}(D\tilde{P}_0^*) - 2\tilde{K}_0^*(DP_{12}) - (K_{10}+K_{10}^*)(D\tilde{P}_0) \Big\} \\ &+\alpha^2 \Big\{ 4\tilde{K}_0^* P_{12} + \tilde{P}_0^* K_{12} + \tilde{P}_0 (K_{10}+K_{10}^*) \Big\} \Big) \hat{u}_P \\ &+\alpha^2 \left(\hat{\theta} - \hat{\phi}\right) \Big\{ 2\theta_{12}(D\tilde{P}_0^*) + 2\tilde{\theta}_0^*(DP_{12}) + \tilde{P}_0^*(D\theta_{12}) \\ &+4P_{12}(D\tilde{\theta}_0^*) + \tilde{P}_0 D(\theta_{10}+\theta_{10}^*) \Big\} - \hat{\phi} D \Big\{ \tilde{\theta}_0^* \phi_{12} \\ &+ \tilde{\theta}_0(\phi_{10}+\phi_{10}^*) + \tilde{\phi}_0^* \theta_{12} + \tilde{\phi}_0(\theta_{10}+\theta_{10}^*) \Big\} \Big] dz \end{aligned} \tag{42}$$

respectively. Now, the equation satisfied by $A(t_s)$ as

$$\frac{dA}{dt_s} = bA + cA|A|^2, \quad \text{with} \quad b = \frac{T_2}{T_1}, \quad c = \frac{T_3}{T_1} \tag{43}$$

where T_1, T_2 and T_3 are respectively given by (40), (41) and (42).

Conjugate of the above equation yields

$$\frac{dA^*}{dt_s} = b^*A^* + c^*A^*|A|^2 \tag{44}$$

Multiplying equation (43) by A^* and equation (44) by A and adding, we obtain Landau-type equation

$$\frac{d\,|A|^2}{dt_s} = 2b_r|A|^2 + 2c_r|A|^4 \tag{45}$$

where b_r and c_r are real parts of b and c respectively. Here $-2c_r$ is the Landau constant. Dividing (45) by $|A|^4$, we get a first-order linear equation in $|A|^{-2}$

$$\frac{d|A|^{-2}}{dt_s} + 2b_r|A|^{-2} \;=\; -2c_r \tag{46}$$

with integrating factor $e^{2b_r t_s}$. Solution can be expressed as

$$|A|^{-2} = \left(A_0^{-2} + \frac{c_r}{b_r}\right) e^{-2b_r t_s} - \frac{c_r}{b_r}$$

if $b_r \neq 0$ and A_0 is the initial value of $|A|$. The above result yields

$$|A|^{-2} = A_0^{-2}\left[\left(1 + \frac{c_r}{b_r}A_0^2\right) e^{-2b_r t_s} - \frac{c_r}{b_r}A_0^2\right]$$

Thus,

$$|A|^2 = \frac{A_0^2}{\left(1 + \frac{c_r}{b_r}A_0^2\right) e^{-2b_r t_s} - \frac{c_r}{b_r}A_0^2} \tag{47}$$

2.6. Numerical Results for Permeable Mush-Liquid Interface

For our computations, we chose the parameters $\mathcal{S} = 3.2$, $\mathcal{C} = 9.0$, $\theta_\infty = 0.1$. Since we take $K = (1-\phi)^{-n}$, we use $n = 3$ for variable permeability case and $n = 0$ for constant permeability case for our computational results. First we obtain the basic state solutions by performing computation on the equations (6) and (7). After computing the basic state solutions, we obtain marginal stability curve which consists of points given by given by the critical Rayleigh number as a function of the wavenumber.These points are obtained by solving the equations (15), (16), and (17) with boundary conditions $\tilde{P}_0 = \tilde{\theta}_0 = 0$ at $z = 0$ and $D\tilde{P}_0 = \tilde{\theta}_0 = \tilde{\phi}_0 = 0$ at $z = \delta$ using a fourth-order Runge-Kutta shooting method. The critical Rayleigh number designated by R_c is the minimum value of R_0 below which no motion can exist based on the linear system. The critical wavenumber, α_c, is the wavenumber that corresponds to the pair $(\alpha_c,\ R_c)$ with the minimum Rayleigh number. This process yields the critical wavenumber and the critical Rayleigh number as 1.35 and 17.6477, respectively, for variable permeability case. For constant permeability case, the critical pair is given by (1.33, 15.6162). Once we obtain the the critical pair $(\alpha_c,\ R_c)$, we solve the linear system (of order ϵ^1) given by (18) for the dependent variables numerically using this critical pair. Then we numerically compute the solutions of the adjoint system given by (22), first-order systems (of order ϵ^2) given by (27) to (32) using a fourth-order Runge-Kutta method. After obtaining the linear, adjoint and first-order systems, we compute the coefficients of Landau-type equation (45) with the expressions (40), (41), and (42) using Simpson's rule. These constants are found to be $b_r = 0.206R_2$ and $c_r = -3.317$ for variable permeability case and $b_r = 0.2308R_2$ and $c_r = -0.8876$ for constant permeability case. Since Landau constant is $-2c_r$, these results indicate that Landau constant has positive sign for either variable or constant permeability. Using Landau theory, we can comment that there is a slow transition to turbulence.

Since the poloidal function is proportional to the vertical component W of the volume flux, the results presented in the figures 2-3 indicate that the time evolution of the flow tends to make both linear and non-linear contributions for W mostly less non-uniform in the vertical direction in the active mush case.

Linear and first-order solid volume fractions are shown in figure 4 and figure 5 for active layer. The results presented in the figures 4-5 for the variable permeability case indicate that as time increases there is less tendency for chimney formation in the linear regime, while there is more tendency for chimney formation near the lower boundary in the non-linear regime.

Variations of the amplitude function A with respect to time presented in the figures 6-7 for variable and constant permeability cases indicate that the flow solution in the active layer eventually reaches a steady state with smaller amplitude, whereas for the passive layer case the flow reaches a steady state with higher amplitude.

3. Mushy Layer with a Magnetic Field

Governing equations for a mushy layer system without magnetic field has been presented in previous section. The non-dimensional form of the equations with the inclusion of magnetic field [3] in the mushy layer are given by

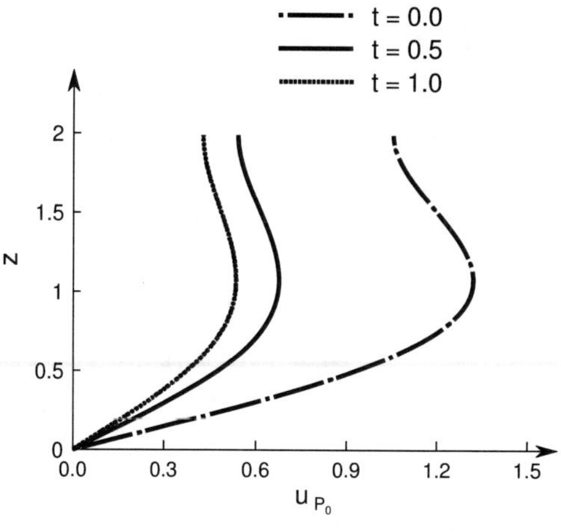

Figure 2. The evolution of the $O(\epsilon^1)$ contribution to the polodial component for the problem with variable permeability, $n = 3$.

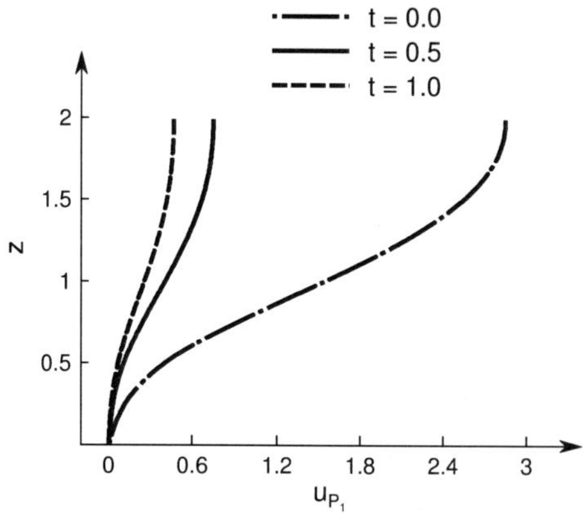

Figure 3. The evolution of the $O(\epsilon^2)$ contribution to the polodial component for the problem with variable permeability, $n = 3$.

$$\left(\frac{\partial}{\partial t} - \frac{\partial}{\partial z}\right)[\Theta - \mathcal{S}\Phi] + \left(\vec{\mathcal{U}}.\nabla\right)\Theta = \nabla^2\Theta \tag{48}$$

$$\left(\frac{\partial}{\partial t} - \frac{\partial}{\partial z}\right)[(1-\Phi)\Theta + \mathcal{C}\Phi] + \left(\vec{\mathcal{U}}.\nabla\right)\Theta = 0 \tag{49}$$

$$\mathcal{K}\vec{\mathcal{U}} + \nabla\mathcal{P} + \mathcal{R}\Theta\hat{k} = \frac{Q}{\tau}\left(\vec{\mathcal{H}}.\nabla\right)\vec{\mathcal{H}} \tag{50}$$

$$\left(\frac{\partial}{\partial t} - \frac{\partial}{\partial z}\right)\vec{\mathcal{H}} + \left(\vec{\mathcal{U}}.\nabla\right)\vec{\mathcal{H}} = \left(\vec{\mathcal{H}}.\nabla\right)\vec{\mathcal{U}} + \frac{1}{\tau}\nabla^2\vec{\mathcal{H}} \tag{51}$$

$$\nabla.\vec{\mathcal{U}} = 0 \tag{52}$$

with boundary conditions:

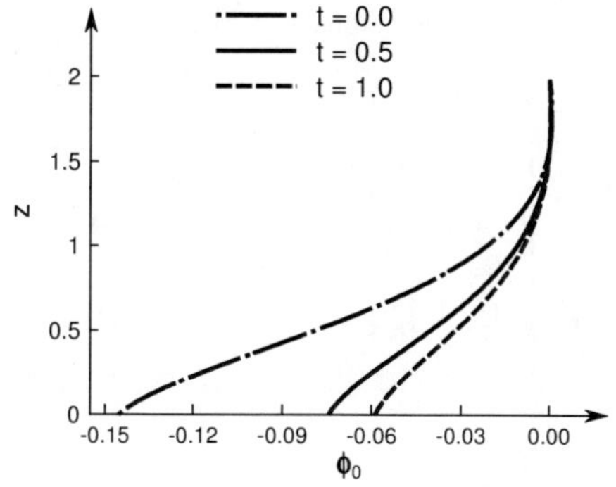

Figure 4. The evolution of the $O(\epsilon^1)$ contribution to the solid volume fraction for the problem with variable permeability, $n = 3$.

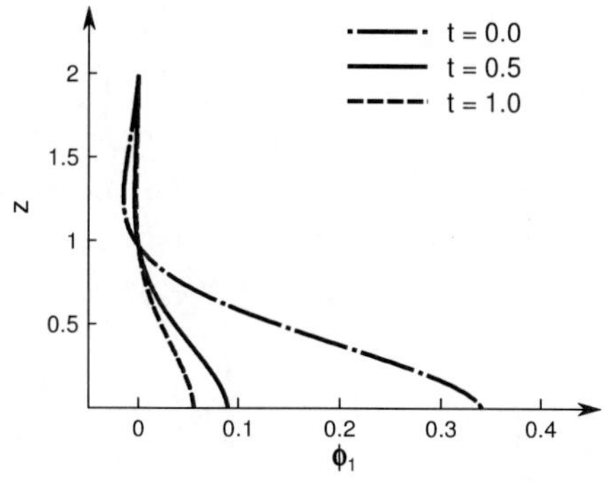

Figure 5. The evolution of the $O(\epsilon^2)$ contribution to the solid volume fraction for the problem with variable permeability, $n = 3$.

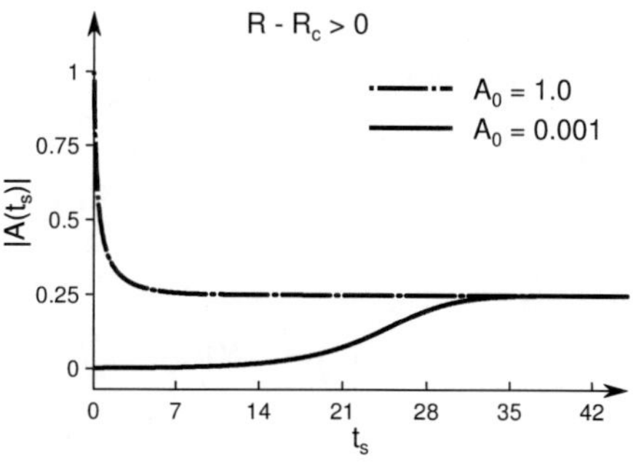

Figure 6. Time evolution of the amplitude for variable permeability $(n = 3)$ for $R - R_c > 0$.

$$\Theta = -1,\ \mathcal{W} = 0,\ \overrightarrow{\mathcal{H}} = \hat{k} \qquad \text{at} \quad z = 0$$
$$\Theta = \Phi = \mathcal{W} = 0,\ \overrightarrow{\mathcal{H}} = \hat{k} \qquad \text{at} \quad z = \delta$$

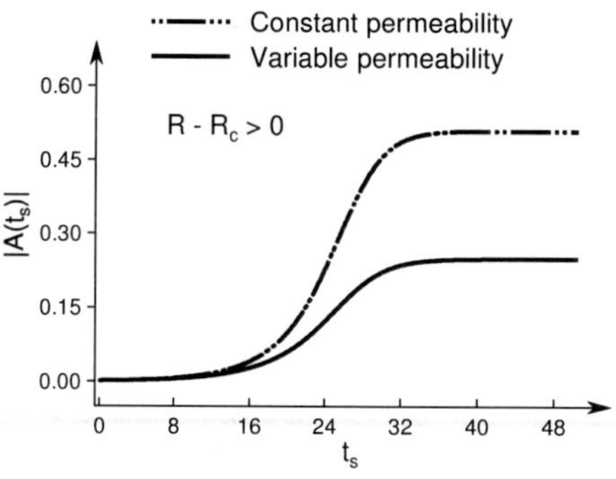

Figure 7. Time evolution of the amplitude for constant permeability ($n = 0$) and variable permeability ($n = 3$) with $A_0 = 0.001$.

Equations (48) and (49) represent conservation of heat and conservation of solute respectively. Darcy law yields equation (50). Equation (51) is known as the magnetic induction equation whereas equation (52) represents conservation of mass. Here Θ, Φ, $\overrightarrow{\mathcal{U}}$, $\mathcal{P}$ and $\overrightarrow{\mathcal{H}}$ are the non dimensional dependent variables representing heat, solid fraction, liquid flux, pressure and magnetic field. Non dimensional numbers appearing in the model are Stefan number $\mathcal{S}$, concentration ratio $\mathcal{C}$, Rayleigh number $\mathcal{R}$, Chandrasekhar number Q and Robert's number τ ([18]). The symbol δ is used to denote nondimensional thickness of the layer. For active layer, we assume that permeability of layer is varying, i.e., in particular, we use $\mathcal{K} = \frac{1}{\chi^n} = \frac{1}{(1-\Phi)^n}$ where χ stands for the liquid fraction.

3.1. Solution Procedure

Assuming solutions of the form

$$\begin{aligned}
\Theta(x,y,z,t) &= \theta_b(z) + \epsilon\theta\,(x,y,z,t) \\
\Phi(x,y,z,t) &= \phi_b(z) + \epsilon\phi\,(x,y,z,t) \\
\overrightarrow{\mathcal{U}}(x,y,z,t) &= \overrightarrow{0} + \epsilon\,\overrightarrow{u}\,(x,y,z,t) \\
\mathcal{P}(x,y,z,t) &= p_b(z) + \epsilon p\,(x,y,z,t) \\
\mathcal{K}(x,y,z,t) &= k_b(z) + \epsilon K\,(x,y,z,t) \\
\overrightarrow{\mathcal{H}}(x,y,z,t) &= \hat{k} + \epsilon\,\overrightarrow{h}\,(x,y,z,t)
\end{aligned} \tag{53}$$

where θ_b, ϕ_b, p_b are solutions to the steady basic state system (system with no flow) and k_b can be obtained from ϕ_b. The variables θ, ϕ, $\overrightarrow{u}$, p, K, $\overrightarrow{h}$ are the perturbations for Θ, Φ, $\overrightarrow{\mathcal{U}}$, $\mathcal{P}$, $\mathcal{K}$ and $\overrightarrow{\mathcal{H}}$ respectively. Here ϵ is the perturbation parameter defined in section 3. Basic state solutions have been derived earlier.

Perturbed system is given by

$$\left(\frac{\partial}{\partial t} - \frac{\partial}{\partial z} - \nabla^2\right)\theta - \mathcal{S}\left(\frac{\partial}{\partial t} - \frac{\partial}{\partial z}\right)\phi + w\frac{d\theta_b}{dz} = -\epsilon\,(\overrightarrow{u}.\nabla\theta) \tag{54}$$

$$\left(\frac{\partial}{\partial t}-\frac{\partial}{\partial z}\right)\{(1-\phi_b)\,\theta+(\mathcal{C}-\theta_b)\,\phi\}+w\frac{d\theta_b}{dz}$$
$$=\epsilon\left\{-\overrightarrow{u}.\nabla\theta+\left(\frac{\partial}{\partial t}-\frac{\partial}{\partial z}\right)(\theta\phi)\right\} \tag{55}$$
$$k_b\overrightarrow{u}+\nabla p+\mathcal{R}\theta\hat{k}-\frac{Q}{\tau}\frac{\partial\overrightarrow{h}}{\partial z}=\epsilon\left(\frac{Q}{\tau}\overrightarrow{h}.\nabla\overrightarrow{h}-K\overrightarrow{u}\right) \tag{56}$$
$$\left(\frac{\partial}{\partial t}-\frac{\partial}{\partial z}\right)\overrightarrow{h}-\frac{\partial\overrightarrow{u}}{\partial z}-\frac{1}{\tau}\left(\nabla^2\overrightarrow{h}\right)=\epsilon\left(\overrightarrow{h}.\nabla\overrightarrow{u}-\overrightarrow{u}.\nabla\overrightarrow{h}\right) \tag{57}$$
$$\nabla.\overrightarrow{u}=0 \tag{58}$$

with $\theta=w=h_3=0$ at $z=0$ and $\theta=\phi=w=h_3=0$ at $z=\delta$. Here w and h_3 are vertical components of $\overrightarrow{u}$ and $\overrightarrow{h}$ respectively.

Writing $\overrightarrow{u}=(u,\,v,\,w)=\nabla\times\nabla\times\left(u_P\hat{k}\right)+\nabla\times\left(u_T\hat{k}\right)$ (as $\nabla.\overrightarrow{u}=0$, Chandrasekhar [3]), where u_P and u_T represent poloidal and toroidal components of $\overrightarrow{u}$. Now the third component of $\nabla\times\nabla\times(k_b\overrightarrow{u})$ is given by

$$\begin{aligned}&\frac{\partial^2(k_bu)}{\partial x\partial z}+\frac{\partial^2(k_bv)}{\partial y\partial z}-\frac{\partial^2(k_bw)}{\partial x^2}-\frac{\partial^2(k_bw)}{\partial y^2}\\&=\frac{\partial}{\partial z}\left[k_b\left(\frac{\partial u}{\partial x}+\frac{\partial v}{\partial y}\right)\right]-k_b\frac{\partial^2w}{\partial x^2}-k_b\frac{\partial^2w}{\partial y^2}\\&=-k_b'\frac{\partial w}{\partial z}-k_b\frac{\partial^2w}{\partial z^2}-k_b\frac{\partial^2w}{\partial x^2}-k_b\frac{\partial^2w}{\partial y^2}\\&=k_b'\frac{\partial}{\partial z}(\triangle_2u_P)+k_b\nabla^2(\triangle_2u_P)\end{aligned}$$

Similarly, for the third component of $\nabla\times\nabla\times(\mathcal{R}\theta\hat{k})$, we have

$$-\mathcal{R}\left[\frac{\partial^2\theta}{\partial x^2}+\frac{\partial^2\theta}{\partial y^2}\right]=-\mathcal{R}(\Delta_2\theta).$$

Writing $\overrightarrow{h}=(h_1,\,h_2,\,h_3)=\nabla\times\nabla\times\left(h_P\hat{k}\right)+\nabla\times\left(h_T\hat{k}\right)$ (as $\nabla.\overrightarrow{h}=0$), we can express

$$\overrightarrow{h}=(h_1,\,h_2,\,h_3)\quad=\quad\left(\frac{\partial^2h_P}{\partial x\partial z}+\frac{\partial h_T}{\partial y},\;\frac{\partial^2h_P}{\partial y\partial z}-\frac{\partial h_T}{\partial x},\;-\Delta_2h_P\right)$$

Now the third component of $\nabla\times\nabla\times\left(\frac{\partial\overrightarrow{h}}{\partial z}\right)$ as

$$\begin{aligned}&\frac{\partial^2}{\partial x\partial z}\left(\frac{\partial h_1}{\partial z}\right)+\frac{\partial^2}{\partial y\partial z}\left(\frac{\partial h_2}{\partial z}\right)-\frac{\partial^2}{\partial x^2}\left(\frac{\partial h_3}{\partial z}\right)-\frac{\partial^2}{\partial y^2}\left(\frac{\partial h_3}{\partial z}\right)\\&=-\frac{\partial^2}{\partial x^2}\left(\frac{\partial h_3}{\partial z}\right)-\frac{\partial^2}{\partial y^2}\left(\frac{\partial h_3}{\partial z}\right)-\frac{\partial^2}{\partial z^2}\left(\frac{\partial h_3}{\partial z}\right)\\&=-\nabla^2\left(\frac{\partial h_3}{\partial z}\right)=\nabla^2\left\{\frac{\partial}{\partial z}\left(\Delta_2h_p\right)\right\}\end{aligned}$$

Thus, the third component of $\nabla \times \nabla \times \left(\frac{Q}{\tau}\frac{\partial \overrightarrow{h}}{\partial z}\right)$ is $\frac{Q}{\tau}\nabla^2\left\{\frac{\partial}{\partial z}\left(\Delta_2 h_p\right)\right\}$.

The third component of $\nabla \times \nabla \times \left(\frac{Q}{\tau}\overrightarrow{h}.\nabla\overrightarrow{h}\right)$ as

$$2\frac{Q}{\tau}\frac{\partial}{\partial z}\left[\left(\frac{\partial}{\partial z}\left(\Delta_2 h_P\right)\right)^2 - \frac{\partial}{\partial x}\left(\frac{\partial^2 h_P}{\partial x \partial z} + \frac{\partial h_T}{\partial y}\right)\frac{\partial}{\partial y}\left(\frac{\partial^2 h_P}{\partial y \partial z} - \frac{\partial h_T}{\partial x}\right)\right.$$
$$+\frac{\partial}{\partial x}\left(\frac{\partial^2 h_P}{\partial y \partial z} - \frac{\partial h_T}{\partial x}\right)\frac{\partial}{\partial y}\left(\frac{\partial^2 h_P}{\partial x \partial z} + \frac{\partial h_T}{\partial y}\right) - \frac{\partial}{\partial x}\left(\Delta_2 h_P\right)$$
$$\left.\times\frac{\partial}{\partial z}\left(\frac{\partial^2 h_P}{\partial x \partial z} + \frac{\partial h_T}{\partial y}\right) - \frac{\partial}{\partial y}\left(\Delta_2 h_P\right)\frac{\partial}{\partial z}\left(\frac{\partial^2 h_P}{\partial y \partial z} - \frac{\partial h_T}{\partial x}\right)\right]$$
$$+\nabla^2\left[\frac{\partial}{\partial x}\left(\Delta_2 h_P\right)\left(\frac{\partial^2 h_P}{\partial x \partial z} + \frac{\partial h_T}{\partial y}\right) + \frac{\partial}{\partial y}\left(\Delta_2 h_P\right)\right.$$
$$\left.\times\left(\frac{\partial^2 h_P}{\partial y \partial z} - \frac{\partial h_T}{\partial x}\right) - \left(\Delta_2 h_P\right)\frac{\partial}{\partial z}\left(\Delta_2 h_P\right)\right]$$

Now the third component of $\nabla \times \nabla \times (-K\overrightarrow{u})$ is given by

$$-\frac{\partial^2}{\partial x \partial z}\left\{K\left(\frac{\partial^2 u_P}{\partial x \partial z} + \frac{\partial u_T}{\partial y}\right)\right\} - \frac{\partial^2}{\partial y \partial z}\left\{K\left(\frac{\partial^2 u_P}{\partial y \partial z} - \frac{\partial u_T}{\partial x}\right)\right\}$$
$$-\frac{\partial^2}{\partial x^2}\left\{K\left(\Delta_2 u_P\right)\right\} - \frac{\partial^2}{\partial y^2}\left\{K\left(\Delta_2 u_P\right)\right\}$$

The third equation of the perturbed system, i.e., equation (56) becomes

$$k_b'\frac{\partial}{\partial z}\left(\Delta_2 u_P\right) + k_b\nabla^2\left(\Delta_2 u_P\right) - \mathcal{R}\left(\Delta_2\theta\right) - \frac{Q}{\tau}\nabla^2\left\{\frac{\partial}{\partial z}\left(\Delta_2 h_P\right)\right\}$$
$$= \frac{\epsilon Q}{\tau}\left[2\frac{\partial}{\partial z}\left\{\left(\frac{\partial}{\partial z}\left(\Delta_2 h_P\right)\right)^2 - \frac{\partial}{\partial x}\left(\frac{\partial^2 h_P}{\partial x \partial z} + \frac{\partial h_T}{\partial y}\right)\frac{\partial}{\partial y}\left(\frac{\partial^2 h_P}{\partial y \partial z} - \frac{\partial h_T}{\partial x}\right)\right.\right.$$
$$+\frac{\partial}{\partial x}\left(\frac{\partial^2 h_P}{\partial y \partial z} - \frac{\partial h_T}{\partial x}\right)\frac{\partial}{\partial y}\left(\frac{\partial^2 h_P}{\partial x \partial z} + \frac{\partial h_T}{\partial y}\right) - \frac{\partial}{\partial z}\left(\frac{\partial^2 h_P}{\partial x \partial z} + \frac{\partial h_T}{\partial y}\right)\frac{\partial}{\partial x}\left(\Delta_2 h_P\right)$$
$$\left.-\frac{\partial}{\partial z}\left(\frac{\partial^2 h_P}{\partial y \partial z} - \frac{\partial h_T}{\partial x}\right)\frac{\partial}{\partial y}\left(\Delta_2 h_P\right)\right\} + \nabla^2\left\{\left(\frac{\partial^2 h_P}{\partial x \partial z} + \frac{\partial h_T}{\partial y}\right)\frac{\partial}{\partial x}\left(\Delta_2 h_P\right)\right.$$
$$\left.\left.+\left(\frac{\partial^2 h_P}{\partial y \partial z} - \frac{\partial h_T}{\partial x}\right)\frac{\partial}{\partial y}\left(\Delta_2 h_P\right) - \left(\Delta_2 h_P\right)\frac{\partial}{\partial z}\left(\Delta_2 h_P\right)\right\}\right]$$
$$-\epsilon\left[\frac{\partial^2}{\partial x \partial z}\left\{K\left(\frac{\partial^2 u_P}{\partial x \partial z} + \frac{\partial u_T}{\partial y}\right)\right\} + \frac{\partial^2}{\partial y \partial z}\left\{K\left(\frac{\partial^2 u_P}{\partial y \partial z} - \frac{\partial u_T}{\partial x}\right)\right\}\right.$$
$$\left.+\frac{\partial^2}{\partial x^2}\left\{K\left(\Delta_2 u_P\right)\right\} + \frac{\partial^2}{\partial y^2}\left\{K\left(\Delta_2 u_P\right)\right\}\right] \tag{59}$$

The first equation of the perturbed system, i.e., equation (54) becomes:

$$\left(\nabla^2 + \frac{\partial}{\partial z} - \frac{\partial}{\partial t}\right)\theta - \mathcal{S}\left(\frac{\partial}{\partial z} - \frac{\partial}{\partial t}\right)\phi + \left(\Delta_2 u_P\right)\frac{d\theta_b}{dz}$$
$$= \epsilon\left[\left(\frac{\partial^2 u_P}{\partial x \partial z} + \frac{\partial u_T}{\partial y}\right)\frac{\partial \theta}{\partial x} + \left(\frac{\partial^2 u_P}{\partial y \partial z} - \frac{\partial u_T}{\partial x}\right)\frac{\partial \theta}{\partial y} - \left(\Delta_2 u_P\right)\frac{\partial \theta}{\partial z}\right] \tag{60}$$

The second equation of the perturbed system, i.e., equation (55) is given by

$$\left(\frac{\partial}{\partial z}-\frac{\partial}{\partial t}\right)\left[\left(\theta_b-\mathcal{C}\right)\phi-\left(1-\phi_b\right)\theta\right]-\left(\Delta_2 u_P\right)\frac{d\theta_b}{dz}$$
$$=-\epsilon\left[\left(\frac{\partial}{\partial z}-\frac{\partial}{\partial t}\right)[\theta\phi]+\left(\frac{\partial^2 u_P}{\partial x\partial z}+\frac{\partial u_T}{\partial y}\right)\frac{\partial\theta}{\partial x}\right.$$
$$\left.+\left(\frac{\partial^2 u_P}{\partial y\partial z}-\frac{\partial u_T}{\partial x}\right)\frac{\partial\theta}{\partial y}-\left(\Delta_2 u_P\right)\frac{\partial\theta}{\partial z}\right] \quad (61)$$

The fourth equation of the perturbed system, i.e., equation (57) yields

$$\left(\frac{\partial}{\partial t}-\frac{\partial}{\partial z}\right)\left(\Delta_2 h_p\right)-\frac{\partial}{\partial z}\left(\Delta_2 u_p\right)-\frac{1}{\tau}\nabla^2\left(\Delta_2 h_p\right)$$
$$=\epsilon\left[\left(\frac{\partial^2 h_P}{\partial x\partial z}+\frac{\partial h_T}{\partial y}\right)\frac{\partial}{\partial x}\left(\Delta_2 u_p\right)+\left(\frac{\partial^2 h_P}{\partial y\partial z}-\frac{\partial h_T}{\partial x}\right)\frac{\partial}{\partial y}\left(\Delta_2 u_p\right)\right.$$
$$+\left(\Delta_2 u_P\right)\frac{\partial}{\partial z}\left(\Delta_2 h_P\right)-\left(\frac{\partial^2 u_P}{\partial x\partial z}+\frac{\partial u_T}{\partial y}\right)\frac{\partial}{\partial x}\left(\Delta_2 h_P\right)$$
$$\left.-\left(\frac{\partial^2 u_P}{\partial y\partial z}-\frac{\partial u_T}{\partial x}\right)\frac{\partial}{\partial y}\left(\Delta_2 h_p\right)-\left(\Delta_2 h_P\right)\frac{\partial}{\partial z}\left(\Delta_2 u_P\right)\right] \quad (62)$$

Now for two-dimensional case, equations (59-62) respectively become

$$k_b'\frac{\partial}{\partial z}\left(\Delta_2 u_P\right)+k_b\nabla^2\left(\Delta_2 u_P\right)-\mathcal{R}\left(\Delta_2\theta\right)-\frac{Q}{\tau}\nabla^2\left\{\frac{\partial}{\partial z}\left(\Delta_2 h_p\right)\right\}$$
$$=\frac{\epsilon Q}{\tau}\left[2\frac{\partial}{\partial z}\left\{\left(\frac{\partial}{\partial z}\left(\Delta_2 h_P\right)\right)^2-\frac{\partial^3 h_P}{\partial x\partial z^2}\frac{\partial}{\partial x}\left(\Delta_2 h_P\right)\right\}\right.$$
$$\left.+\nabla^2\left\{\frac{\partial^2 h_P}{\partial x\partial z}\frac{\partial}{\partial x}\left(\Delta_2 h_P\right)-\left(\Delta_2 h_P\right)\frac{\partial}{\partial z}\left(\Delta_2 h_P\right)\right\}\right]$$
$$-\epsilon\left[\frac{\partial^2}{\partial x\partial z}\left\{K\left(\frac{\partial^2 u_P}{\partial x\partial z}\right)\right\}+\frac{\partial^2}{\partial x^2}\left\{K\left(\Delta_2 u_P\right)\right\}\right] \quad (63)$$

$$\left(\nabla^2+\frac{\partial}{\partial z}-\frac{\partial}{\partial t}\right)\theta-\mathcal{S}\left(\frac{\partial}{\partial z}-\frac{\partial}{\partial t}\right)\phi+\left(\Delta_2 u_P\right)\frac{d\theta_b}{dz}$$
$$=\epsilon\left[\frac{\partial^2 u_P}{\partial x\partial z}\frac{\partial\theta}{\partial x}-\left(\Delta_2 u_P\right)\frac{\partial\theta}{\partial z}\right] \quad (64)$$

$$\left(\frac{\partial}{\partial z}-\frac{\partial}{\partial t}\right)\left[\left(\theta_b-\mathcal{C}\right)\phi-\left(1-\phi_b\right)\theta\right]-\left(\Delta_2 u_P\right)\frac{d\theta_b}{dz}$$
$$=-\epsilon\left[\left(\frac{\partial}{\partial z}-\frac{\partial}{\partial t}\right)[\theta\phi]+\frac{\partial^2 u_P}{\partial x\partial z}\frac{\partial\theta}{\partial x}-\left(\Delta_2 u_P\right)\frac{\partial\theta}{\partial z}\right] \quad (65)$$

and

$$\left(\frac{\partial}{\partial t}-\frac{\partial}{\partial z}\right)(\Delta_2 h_p)-\frac{\partial}{\partial z}(\Delta_2 u_p)-\frac{1}{\tau}\nabla^2(\Delta_2 h_p)$$
$$=\epsilon\left[\frac{\partial^2 h_P}{\partial x\partial z}\frac{\partial}{\partial x}(\Delta_2 u_p)+(\Delta_2 u_P)\frac{\partial}{\partial z}(\Delta_2 h_P)\right.$$
$$\left.-\frac{\partial^2 u_P}{\partial x\partial z}\frac{\partial}{\partial x}(\Delta_2 h_P)-(\Delta_2 h_P)\frac{\partial}{\partial z}(\Delta_2 u_P)\right] \quad (66)$$

3.2. Higher Order Systems

Now we consider linear, first-order and second-order solutions by writing as follows

$$\begin{aligned}
u_P(x,z,t) &= u_{P_0}(x,z,t)+\epsilon u_{P_1}(x,z,t)+\epsilon^2 u_{P_2}(x,z,t)+\ldots..\\
\theta(x,z,t) &= \theta_0(x,z,t)+\epsilon\theta_1(x,z,t)+\epsilon^2\theta_2(x,z,t)+\ldots..\\
\phi(x,z,t) &= \phi_0(x,z,t)+\epsilon\phi_1(x,z,t)+\epsilon^2\phi_2(x,z,t)+\ldots.. \qquad (67)\\
h_P(x,z,t) &= h_{P_0}(x,z,t)+\epsilon h_{P_1}(x,z,t)+\epsilon^2 h_{P_2}(x,z,t)+\ldots..
\end{aligned}$$

Following Drazin and Reid [2], we define slow time variable t_s given by

$$t_s=\epsilon^2 t \qquad \text{with} \qquad \epsilon^2=\frac{\mathcal{R}-\mathcal{R}_c}{\mathcal{R}_2}.$$

Here $\mathcal{R}_c$ is the critical Rayleigh number and the sign and magnitude of $\mathcal{R}_2$ is chosen so that ϵ^2 remains small and positive.

3.2.1. Linear System

Linear system can be expressed as

$$\begin{aligned}
k_b'\frac{\partial}{\partial z}(\Delta_2 u_{P_0})+k_b\nabla^2(\Delta_2 u_{P_0})-\mathcal{R}_c(\Delta_2\theta_0)-\frac{Q}{\tau}\nabla^2\left\{\frac{\partial}{\partial z}(\Delta_2 h_{P_0})\right\} &= 0\\
\left(\nabla^2+\frac{\partial}{\partial z}-\frac{\partial}{\partial t}\right)\theta_0-\mathcal{S}\left(\frac{\partial}{\partial z}-\frac{\partial}{\partial t}\right)\phi_0+(\Delta_2 u_{P_0})\frac{d\theta_b}{dz} &= 0\\
\left(\frac{\partial}{\partial z}-\frac{\partial}{\partial t}\right)[(1-\phi_b)\theta_0+(\mathcal{C}-\theta_b)\phi_0]+(\Delta_2 u_{P_0})\frac{d\theta_b}{dz} &= 0 \quad (68)\\
\left(\frac{\partial}{\partial z}-\frac{\partial}{\partial t}\right)(\Delta_2 h_{P_0})+\frac{\partial}{\partial z}(\Delta_2 u_{P_0})+\frac{1}{\tau}\nabla^2(\Delta_2 h_{P_0}) &= 0
\end{aligned}$$

with boundary conditions:

$$\begin{aligned}
u_{P_0}=\theta_0=h_{P_0}=0 &\qquad \text{at} \quad z=0\\
u_{P_0}=\theta_0=\phi_0=h_{P_0}=0 &\qquad \text{at} \quad z=\delta.
\end{aligned}$$

Using $\frac{\partial}{\partial t} \to \frac{\partial}{\partial t} + \epsilon^2 \frac{\partial}{\partial t_s}$ and considering slow time only together with the assumption of normal mode solutions ([2], [3]), i.e., solutions are proportional to $e^{i\alpha x}$, i.e.,

$$\begin{aligned} u_{P_0}(x,z,t_s) &= A(t_s)\,\tilde{u_{P_0}}(z)e^{i\alpha x} + C.C. \\ \theta_0(x,z,t_s) &= A(t_s)\tilde{\theta_0}(z)e^{i\alpha x} + C.C. \\ \phi_0(x,z,t_s) &= A(t_s)\tilde{\phi_0}(z)e^{i\alpha x} + C.C. \\ h_{P_0}(x,z,t_s) &= A(t_s)\,\tilde{h_{P_0}}(z)e^{i\alpha x} + C.C. \end{aligned}$$

the linear system can be obtained as

$$\left[(Q+k_b)\,D^2 + k_b'D - \alpha^2 k_b\right]\tilde{u_{P_0}} - \mathcal{R}_c\tilde{\theta_0} + Q\left(D^2\,\tilde{h_{P_0}}\right) = 0 \tag{69}$$

$$-\alpha^2\theta_b'\,\tilde{u_{P_0}} + \left[D^2 + D - \alpha^2\right]\tilde{\theta_0} - \mathcal{S}\left(D\tilde{\phi_0}\right) = 0 \tag{70}$$

$$\alpha^2\theta_b'\,\tilde{u_{P_0}} + \left[\phi_b' - (1-\phi_b)\,D\right]\tilde{\theta_0} + \left[\theta_b' - (\mathcal{C}-\theta_b)\,D\right]\tilde{\phi_0} = 0 \tag{71}$$

$$\tau D\,\tilde{u_{P_0}} + \left[D^2 + \tau D - \alpha^2\right]\tilde{h_{P_0}} = 0 \tag{72}$$

where $D = \frac{d}{dz}$.

Thus the linear system can be written as

$$Lq = 0 \tag{73}$$

with L as

$$\begin{bmatrix} (Q+k_b)\,D^2 + k_b'D - \alpha^2 k_b & -\mathcal{R}_c & 0 & QD^2 \\ -\alpha^2\theta_b' & D^2 + D - \alpha^2 & -\mathcal{S}D & 0 \\ \alpha^2\theta_b' & \phi_b' - (1-\phi_b)\,D & \theta_b' - (\mathcal{C}-\theta_b)\,D & 0 \\ \tau D & & & D^2 + \tau D - \alpha^2 \end{bmatrix} \tag{74}$$

where $q = \begin{bmatrix} \tilde{u_{P_0}}, & \tilde{\theta_0}, & \tilde{\phi_0}, & \tilde{h_{P_0}} \end{bmatrix}^T$. Here T in $[.]^T$ is used to denote the transpose.

3.2.2. Adjoint System

Now we define an adjoint operator L_a by

$$< q_a,\, Lq > \; = \; < L_a q_a,\, q > \tag{75}$$

where L and q are given in (74), $q_a = \left[u_{P_a},\, \theta_a,\, \phi_a,\, h_{P_a}\right]^T$ and $< f,\; g > = \int_0^\delta f\,.\,g^* dz$. Since $< f,\; g^* > = < g,\; f^* >$, we have

$$< Lq,\, q_a^* > = < q_a,\, Lq^* > = < L_a q_a,\, q^* >$$

i.e.,

$$\int_0^\delta Lq\,.\,q_a\,dz \quad = \quad \int_0^\delta L_a q_a\,.\,q\,dz$$

To derive the adjoint system, we multiply the equations (69), (70), (71) and (72) by u_{P_a}, θ_a ϕ_a and h_{P_a} respectively, and then integrate with respect to z from $z = 0$ to $z = \delta$ to obtain

$$\int_0^\delta \left\{\left[(Q + k_b)\,D^2 + k_b' D - \alpha^2 k_b\right]\,\tilde{u_{P_0}} - \mathcal{R}_c\tilde{\theta}_0 + Q\left(D^2\,\tilde{h_{P_0}}\right)\right\} u_{P_a} dz$$
$$+ \int_0^\delta \left\{-\alpha^2\theta_b'\,\tilde{u_{P_0}} + \left[D^2 + D - \alpha^2\right]\tilde{\theta}_0 - \mathcal{S}\left(D\tilde{\phi_0}\right)\right\}\theta_a dz$$

$$+ \int_0^\delta \left\{\alpha^2\theta_b'\,\tilde{u_{P_0}} + \left[\phi_b' - (1 - \phi_b)\,D\right]\tilde{\theta}_0 + \left[\theta_b' - (\mathcal{C} - \theta_b)\,D\right]\tilde{\phi_0}\right\}\phi_a dz$$
$$+ \int_0^\delta \left\{\tau D\,\tilde{u_{P_0}} + \left[D^2 + \tau D - \alpha^2\right]\,\tilde{h_{P_0}}\right\} h_{P_a} dz = 0$$

Boundary conditions for adjoint system are

$$u_{P_a} = \theta_a = \phi_a = h_{P_a} = 0 \qquad \text{at} \quad z = 0$$
$$u_{P_a} = \theta_a = h_{P_a} = 0 \qquad \text{at} \quad z = \delta.$$

Integrating by parts and using the boundary conditions, the adjoint system can be expressed as

$$\begin{aligned}
\left[(Q + k_b)\,D^2 + k_b' D - \alpha^2 k_b\right] u_{P_a} + \alpha^2\theta_b'\,(\phi_a - \theta_a) - \tau\,(D h_{P_a}) &= 0 \\
-\mathcal{R}_c u_{P_a} + \left[D^2 - D - \alpha^2\right]\theta_a + (1 - \phi_b)\,(D\phi_a) &= 0 \\
\mathcal{S}\,(D\theta_a) + (\mathcal{C} - \theta_b)\,(D\phi_a) &= 0 \\
Q\left(D^2 u_{P_a}\right) + \left[D^2 - \tau D - \alpha^2\right] h_{P_a} &= 0
\end{aligned} \tag{76}$$

Thus, we obtain adjoint operator L_a as follows

$$\begin{bmatrix}
(Q + k_b)\,D^2 + k_b' D - \alpha^2 k_b & -\alpha^2\theta_b' & \alpha^2\theta_b' & -\tau D \\
-\mathcal{R}_c & D^2 - D - \alpha^2 & (1 - \phi_b)\,D & 0 \\
0 & \mathcal{S}D & (\mathcal{C} - \theta_b)\,D & 0 \\
QD^2 & & & D^2 - \tau D - \alpha^2
\end{bmatrix} \tag{77}$$

3.2.3. First Order System

Equations for the first-order become

$$
\begin{aligned}
&k_b' \frac{\partial}{\partial z}\left(\Delta_2 u_{P_1}\right) + k_b \nabla^2 \left(\Delta_2 u_{P_1}\right) - \mathcal{R}_c \left(\Delta_2 \theta_1\right) - \frac{Q}{\tau} \nabla^2 \left\{ \frac{\partial}{\partial z}\left(\Delta_2 h_{P_1}\right) \right\} \\
&= \frac{Q}{\tau}\left[2\frac{\partial}{\partial z}\left\{ \left(\frac{\partial}{\partial z}\left(\Delta_2 h_{P_0}\right)\right)^2 - \frac{\partial^3 h_{P_0}}{\partial x \partial z^2}\frac{\partial}{\partial x}\left(\Delta_2 h_{P_0}\right)\right\} \right. \\
&\left. + \nabla^2 \left\{ \frac{\partial^2 h_{P_0}}{\partial x \partial z}\frac{\partial}{\partial x}\left(\Delta_2 h_{P_0}\right) - \left(\Delta_2 h_{P_0}\right)\frac{\partial}{\partial z}\left(\Delta_2 h_{P_0}\right)\right\}\right] \\
&- \left[\frac{\partial^2}{\partial x \partial z}\left\{ K_0 \left(\frac{\partial^2 u_{P_0}}{\partial x \partial z}\right)\right\} + \frac{\partial^2}{\partial x^2}\left\{ K_0 \left(\Delta_2 u_{P_0}\right)\right\}\right]
\end{aligned} \tag{78}
$$

$$
\left(\nabla^2 + \frac{\partial}{\partial z}\right)\theta_1 - \mathcal{S}\frac{\partial \phi_1}{\partial z} + \theta_b' \left(\Delta_2 u_{P_1}\right) = \frac{\partial^2 u_{P_0}}{\partial x \partial z}\frac{\partial \theta_0}{\partial x} - \left(\Delta_2 u_{P_0}\right)\frac{\partial \theta_0}{\partial z} \tag{79}
$$

$$
\begin{aligned}
&\frac{\partial}{\partial z}\left[(1-\phi_b)\theta_1 + (\mathcal{C} - \theta_b)\phi_1\right] + \theta_b' \left(\Delta_2 u_{P_1}\right) \\
&= \frac{\partial}{\partial z}\left[\theta_0 \phi_0\right] + \frac{\partial^2 u_{P_0}}{\partial x \partial z}\frac{\partial \theta_0}{\partial x} - \left(\Delta_2 u_{P_0}\right)\frac{\partial \theta_0}{\partial z}
\end{aligned} \tag{80}
$$

and

$$
\begin{aligned}
&\frac{\partial}{\partial z}\left(\Delta_2 h_{P_1}\right) + \frac{\partial}{\partial z}\left(\Delta_2 u_{P_1}\right) + \frac{1}{\tau}\nabla^2 \left(\Delta_2 h_{P_1}\right) \\
&= -\frac{\partial^2 h_{P_0}}{\partial x \partial z}\frac{\partial}{\partial x}\left(\Delta_2 u_{P_0}\right) - \left(\Delta_2 u_{P_0}\right)\frac{\partial}{\partial z}\left(\Delta_2 h_{P_0}\right) \\
&+ \frac{\partial^2 u_{P_0}}{\partial x \partial z}\frac{\partial}{\partial x}\left(\Delta_2 h_{P_0}\right) + \left(\Delta_2 h_{P_0}\right)\frac{\partial}{\partial z}\left(\Delta_2 u_{P_0}\right)
\end{aligned} \tag{81}
$$

Using normal mode approach (Drazin and Reid [2], Chandrasekhar [3]), we write the first-order quantities as

$$
\begin{aligned}
u_{P_1}(x,z,t_s) &= A^2(t_s) u_{P_{12}}(z) e^{2i\alpha x} + |A(t_s)|^2 u_{P_{10}}(z) + C.C. \\
\theta_1(x,z,t_s) &= A^2(t_s) \theta_{12}(z) e^{2i\alpha x} + |A(t_s)|^2 \theta_{10}(z) + C.C. \\
\phi_1(x,z,t_s) &= A^2(t_s) \phi_{12}(z) e^{2i\alpha x} + |A(t_s)|^2 \phi_{10}(z) + C.C. \\
h_{P_1}(x,z,t_s) &= A^2(t_s) h_{P_{12}}(z) e^{2i\alpha x} + |A(t_s)|^2 h_{P_{10}}(z) + C.C.
\end{aligned}
$$

where sub-index 12 is used for first-order two$-\alpha$ mode and 10 is used for first-order zero$-\alpha$ mode.

Two- α Mode

Two-α mode equations corresponding to equations (78), (79), (80) and (81) become

$$\left(k_b D^2 + k_b' D - \alpha^2 k_b\right) u_{P_{12}} - \mathcal{R}_c \theta_{12} - \frac{Q}{\tau} D\left(D^2 - 4\alpha^2\right) h_{P_{12}}$$

$$= \frac{\alpha^2 Q}{2\tau}\left[\tilde{h_{P_0}}\left(D^3 \tilde{h_{P_0}}\right) - \left(D \tilde{h_{P_0}}\right)\left(D^2 \tilde{h_{P_0}}\right)\right] + \alpha^2 \tilde{K}_0 \tilde{u_{P_0}} - \frac{1}{2} D\left\{\tilde{K}_0\left(D \tilde{u_{P_0}}\right)\right\} \quad (82)$$

$$-4\alpha^2 \theta_b' u_{P_{12}} + \left(D^2 + D - 4\alpha^2\right)\theta_{12} - \mathcal{S}\left(D\phi_{12}\right)$$
$$= \alpha^2\left[\tilde{u_{P_0}}\left(D\tilde{\theta_0}\right) - \tilde{\theta_0}\left(D\tilde{u_{P_0}}\right)\right] \quad (83)$$

$$-4\alpha^2 \theta_b' u_{P_{12}} + \left[(1-\phi_b) D - \phi_b'\right]\theta_{12} + D\left[(\mathcal{C} - \theta_b) - \theta_b'\right]\phi_{12}$$
$$= D\left(\tilde{\theta_0}\tilde{\phi_0}\right) + \alpha^2\left[\tilde{u_{P_0}}\left(D\tilde{\theta_0}\right) - \tilde{\theta_0}\left(D\tilde{u_{P_0}}\right)\right] \quad (84)$$

$$\tau D u_{P_{12}} + \left[D^2 + \tau D - 4\alpha^2\right] h_{P_{12}} = \frac{\alpha^2 \tau}{2}\left[\tilde{u_{P_0}}\left(D\tilde{h_{P_0}}\right) - \tilde{h_{P_0}}\left(D\tilde{u_{P_0}}\right)\right] \quad (85)$$

Zero- α Mode

Zero-α mode equations corresponding to equations (78), (79), (80) and (81) become

$$\left[k_b D^2 + k_b' D\right] u_{P_{10}} - \mathcal{R}_c \theta_{10} - \frac{Q}{\tau}\left(D^3 h_{P_{10}}\right) = 0 \quad (86)$$

$$\left(D^2 + D\right)\theta_{10} - \mathcal{S}\left(D\phi_{10}\right) = \alpha^2 D\left[\tilde{u_{P_0}}\left(\tilde{\theta_0}\right)^* + \tilde{\theta_0}\left(\tilde{u_{P_0}}\right)^*\right] \quad (87)$$

$$(1-\phi_b)\left(D\theta_{10}\right) - \phi_b'\theta_{10} + (\mathcal{C} - \theta_b)\left(D\phi_{10}\right) - \theta_b'\phi_{10}$$
$$= D\left[\tilde{\theta_0^*}\tilde{\phi_0} + \tilde{\theta_0}\tilde{\phi_0^*} + \alpha^2\left(\tilde{u_{P_0}}\left(\tilde{\theta_0}\right)^* + \tilde{\theta_0}\left(\tilde{u_{P_0}}\right)^*\right)\right] \quad (88)$$

$$\tau\left(D u_{P_{10}}\right) + \left[D^2 + \tau D\right] h_{P_{10}} = 0 \quad (89)$$

3.2.4. Second Order System

For the second-order, equation (63) becomes

$$k_b' \frac{\partial}{\partial z}\left(\Delta_2 u_{P_2}\right) + k_b \nabla^2\left(\Delta_2 u_{P_2}\right) - \mathcal{R}_c\left(\Delta_2 \theta_2\right) - \frac{Q}{\tau}\nabla^2\left\{\frac{\partial}{\partial z}\left(\Delta_2 h_{p_2}\right)\right\}$$
$$= \mathcal{R}_2\left(\Delta_2\theta_0\right) + \frac{Q}{\tau}\left[2\frac{\partial}{\partial z}\left\{\left(\frac{\partial}{\partial z}\left(\Delta_2 h_{P_0}\right)\right)\left(\frac{\partial}{\partial z}\left(\Delta_2 h_{P_1}\right)\right) - \frac{\partial^3 h_{P_0}}{\partial x \partial z^2}\frac{\partial}{\partial x}\left(\Delta_2 h_{P_1}\right)\right.\right.$$
$$\left.- \frac{\partial^3 h_{P_1}}{\partial x \partial z^2}\frac{\partial}{\partial x}\left(\Delta_2 h_{P_0}\right)\right\} + \nabla^2\left\{\frac{\partial^2 h_{P_0}}{\partial x \partial z}\frac{\partial}{\partial x}\left(\Delta_2 h_{P_1}\right) + \frac{\partial^2 h_{P_1}}{\partial x \partial z}\frac{\partial}{\partial x}\left(\Delta_2 h_{P_0}\right)\right.$$
$$\left.\left. - \left(\Delta_2 h_{P_0}\right)\frac{\partial}{\partial z}\left(\Delta_2 h_{P_1}\right) - \left(\Delta_2 h_{P_1}\right)\frac{\partial}{\partial z}\left(\Delta_2 h_{P_0}\right)\right\}\right]$$
$$- \left[\frac{\partial^2}{\partial x \partial z}\left\{K_0\left(\frac{\partial^2 u_{P_1}}{\partial x \partial z}\right)\right\} + \frac{\partial^2}{\partial x \partial z}\left\{K_1\left(\frac{\partial^2 u_{P_0}}{\partial x \partial z}\right)\right\}\right.$$
$$\left. + \frac{\partial^2}{\partial x^2}\left\{K_0\left(\Delta_2 u_{P_1}\right)\right\} + \frac{\partial^2}{\partial x^2}\left\{K_1\left(\Delta_2 u_{P_0}\right)\right\}\right] \quad (90)$$

Equation (64) gives

$$\left(\nabla^2 + \frac{\partial}{\partial z}\right)\theta_2 - \mathcal{S}\frac{\partial \phi_2}{\partial z} + (\Delta_2 u_{P_2})\,\theta_b' = \frac{\partial}{\partial t_s}(S\phi_0 - \theta_0)$$
$$+\frac{\partial^2 u_{P_0}}{\partial x \partial z}\frac{\partial \theta_1}{\partial x} + \frac{\partial^2 u_{P_1}}{\partial x \partial z}\frac{\partial \theta_0}{\partial x} - (\Delta_2 u_{P_0})\frac{\partial \theta_1}{\partial z} - (\Delta_2 u_{P_1})\frac{\partial \theta_0}{\partial z} \tag{91}$$

Equation (65) yields

$$\frac{\partial}{\partial z}\left[(1-\phi_b)\,\theta_2 + (\mathcal{C} - \theta_b)\,\phi_2\right] + (\Delta_2 u_{P_2})\,\theta_b'$$
$$= \frac{\partial}{\partial t_s}\left[(1-\phi_b)\,\theta_0 + (\mathcal{C} - \theta_b)\,\phi_0\right] - (\Delta_2 u_{P_0})\frac{\partial \theta_1}{\partial z} - (\Delta_2 u_{P_1})\frac{\partial \theta_0}{\partial z}$$
$$+\frac{\partial}{\partial z}(\theta_0\phi_1 + \theta_1\phi_0) + \frac{\partial^2 u_{P_0}}{\partial x \partial z}\frac{\partial \theta_1}{\partial x} + \frac{\partial^2 u_{P_1}}{\partial x \partial z}\frac{\partial \theta_0}{\partial x} \tag{92}$$

Equation (66) yields

$$\frac{\partial}{\partial z}(\Delta_2 h_{p_2}) + \frac{\partial}{\partial z}(\Delta_2 u_{p_2}) + \frac{1}{\tau}\nabla^2(\Delta_2 h_{p_2}) - \frac{\partial h_{p_0}}{\partial t_s}$$
$$= \frac{\partial^2 u_{P_0}}{\partial x \partial z}\frac{\partial}{\partial x}(\Delta_2 h_{P_1}) + \frac{\partial^2 u_{P_1}}{\partial x \partial z}\frac{\partial}{\partial x}(\Delta_2 h_{P_0}) - (\Delta_2 u_{P_0})\frac{\partial}{\partial z}(\Delta_2 h_{P_1})$$
$$-(\Delta_2 u_{P_1})\frac{\partial}{\partial z}(\Delta_2 h_{P_0}) - \frac{\partial^2 h_{P_0}}{\partial x \partial z}\frac{\partial}{\partial x}(\Delta_2 u_{P_1}) - \frac{\partial^2 h_{P_1}}{\partial x \partial z}\frac{\partial}{\partial x}(\Delta_2 u_{P_0})$$
$$+(\Delta_2 h_{P_1})\frac{\partial}{\partial z}(\Delta_2 u_{P_0}) + (\Delta_2 h_{P_1})\frac{\partial}{\partial z}(\Delta_2 u_{P_0}) \tag{93}$$

Using normal mode approach, the second order system is obtained as presented below. The first equation of the second order system becomes

$$\left\{(Q + k_b)\,D^2 + k_b' D - \alpha^2 k_b\right\}\tilde{u_{P_2}} - \mathcal{R}_c\tilde{\theta_2} + Q\left(D^2\,\tilde{h_{P_2}}\right)$$
$$= A\mathcal{R}_2\tilde{\theta_0} + Q\left(D\,\tilde{h_{P_2}}\right)\frac{dA}{dt_s} + |A|^2 A\left[\alpha^2\left\{4\tilde{K_0^*}u_{P_{12}} + K_{12}\tilde{u}^*_{P_0}\right.\right.$$
$$\left.+\,(K_{10} + K_{10}^*)\,\tilde{u}_{P_0}\right\} + D\left\{K_{12}\left(D\tilde{u}^*_{P_0}\right) - 2\tilde{K_0^*}\left(Du_{P_{12}}\right) + (K_{10} + K_{10}^*)\left(D\tilde{u}_{P_0}\right)\right\}$$
$$+2\alpha^2 Q\left\{D\left[2h_{P_{12}}\left(D\tilde{u}^*_{P_0}\right) - \tilde{h}^*_{P_0}\left(Du_{P_{12}}\right) + \tilde{u}^*_{P_0}\left(Dh_{P_{12}}\right) - 2u_{12}\left(D\tilde{h}^*_{P_0}\right)\right]\right.$$
$$+\frac{1}{\tau}\left(D\left[\left(D\tilde{h}^*_{P_0}\right)\left(Dh_{P_{12}}\right) - 2h_{P_{12}}\left(D^2\tilde{h}^*_{P_0}\right) + \tilde{h}^*_{P_0}\left(D^2 h_{P_{12}}\right)\right]\right.$$
$$\left.\left.\left.-3\alpha^2\left[2h_{P_{12}}\left(D\tilde{h}^*_{P_0}\right) + \tilde{h}^*_{P_0}\left(Dh_{P_{12}}\right)\right]\right)\right\}\right] \tag{94}$$

From equation (91), we obtain

$$
\begin{aligned}
&-\alpha^2\,\tilde{u_{P_2}}\theta_b' + \left(D^2 + D - \alpha^2\right)\tilde{\theta}_2 - \mathcal{S}\left(D\tilde{\phi}_2\right)\\
&= -\left(\mathcal{S}\tilde{\phi}_0 - \tilde{\theta}_0\right)\frac{dA}{dt_s} + \alpha^2|A|^2A\left\{D\left(2\,\tilde{\theta}_0^* u_{P_{12}} + \tilde{u_{P_0}}^*\theta_{12}\right)\right.\\
&\quad\left. +2u_{P_{12}}(D\tilde{\theta}_0^*) + \theta_{12}(D\,\tilde{u_{P_0}}^*) + \tilde{u_{P_0}}D(\theta_{10} + \theta_{10}^*)\right\} \qquad (95)
\end{aligned}
$$

Equation (92) gives

$$
\begin{aligned}
&\alpha^2\,\tilde{u_{P_2}}\theta_b' - D\left[(1-\phi_b)\,\tilde{\theta}_2 + (\mathcal{C} - \theta_b)\,\tilde{\phi}_2\right]\\
&= -\left[(1-\phi_b)\,\tilde{\theta}_0 + (\mathcal{C} - \theta_b)\,\tilde{\phi}_0\right]\frac{dA}{dt_s} - |A|^2A\left[D\left\{\tilde{\theta}_0^*\phi_{12} + \tilde{\phi}_0^*\theta_{12}\right.\right.\\
&\quad\left. +\tilde{\theta}_0(\phi_{10} + \phi_{10}^*) + \tilde{\phi}_0(\theta_{10} + \theta_{10}^*)\right\} + \alpha^2\left\{D\left(2\,\tilde{\theta}_0^* u_{P_{12}} + \tilde{u_{P_0}}^*\theta_{12}\right)\right.\\
&\quad\left.\left. +2u_{P_{12}}(D\tilde{\theta}_0^*) + \theta_{12}(D\,\tilde{u_{P_0}}^*) + \tilde{u_{P_0}}D(\theta_{10} + \theta_{10}^*)\right\}\right] \qquad (96)
\end{aligned}
$$

and equation (93) becomes

$$
\begin{aligned}
&\tau D\,\tilde{u_{P_2}} + \left(D^2 + \tau D - \alpha^2\right)\,\tilde{h_{P_2}}\\
&= \tau\,\tilde{h_{P_0}}\frac{dA}{dt_s} + 2\alpha^2\tau|A|^2A\left[2h_{P_{12}}\,(D\,\tilde{u_{P_0}}^*) - \tilde{h_{P_0}}^*(Du_{P_{12}})\right.\\
&\quad\left. -2u_{P_{12}}\left(D\,\tilde{h_{P_0}}^*\right) + \tilde{u_{P_0}}^*(Dh_{P_{12}})\right] \qquad (97)
\end{aligned}
$$

3.3. Landau Equation for Magnetic Case

Multiplying (94), (95), (96), (97) by u_{P_a}, θ_a, ϕ_a, h_{P_a}, and integrating, we obtain an equation with LHS and RHS as

$$
\begin{aligned}
\text{LHS} \;=\; & \int_0^\delta u_{P_a}\left[\left\{(Q + k_b)\,D^2 + k_b'D - \alpha^2 k_b\right\}\tilde{u}_{P_2} - \mathcal{R}_c\tilde{\theta}_2 + Q\left(D^2\tilde{h}_{P_2}\right)\right]dz\\
& + \int_0^\delta \theta_a\left[-\alpha^2\tilde{u}_{P_2}\theta_b' + \left(D^2 + D - \alpha^2\right)\tilde{\theta}_2 - \mathcal{S}\left(D\tilde{\phi}_2\right)\right]dz\\
& + \int_0^\delta \phi_a\left[\alpha^2\tilde{u}_{P_2}\theta_b' - D\left\{(1-\phi_b)\,\tilde{\theta}_2 + (\mathcal{C} - \theta_b)\,\tilde{\phi}_2\right\}\right]dz\\
& + \int_0^\delta h_{P_a}\left[\tau D\tilde{u}_{P_2} + \left(D^2 + \tau D - \alpha^2\right)\tilde{h}_{P_2}\right]dz
\end{aligned}
$$

and

$$
\begin{aligned}
\text{RHS} \;=\; & A\mathcal{R}_2\int_0^\delta u_{P_a}\tilde{\theta}_0 dz - \frac{dA}{dt_s}\int_0^\delta \phi_a\left[(1-\phi_b)\,\tilde{\theta}_0 + (\mathcal{C} - \theta_b)\,\tilde{\phi}_0\right]dz\\
& - \frac{dA}{dt_s}\int_0^\delta \theta_a\left(\mathcal{S}\tilde{\phi}_0 - \tilde{\theta}_0\right)dz + Q\frac{dA}{dt_s}\int_0^\delta u_{P_a}\left(D\tilde{h}_{P_0}\right)dz + \tau\frac{dA}{dt_s}\int_0^\delta h_{P_a}\tilde{h}_{P_0}dz
\end{aligned}
$$

$$
\begin{aligned}
&+\alpha^2|A|^2A\int_0^\delta \left[u_{P_a}\frac{2Q}{\tau}\left\{D\left\{(Dh_{P_{12}})\left(D\tilde{h}^*_{P_0}\right)+\tilde{h}^*_{P_0}\left(D^2h_{P_{12}}\right)-2h_{P_{12}}\left(D^2\tilde{h}^*_{P_0}\right)\right\}\right.\right.\\
&+\tau D\left\{2h_{P_{12}}\left(D\tilde{u}^*_{P_0}\right)-\tilde{h}^*_{P_0}\left(Du_{P_{12}}\right)+\tilde{u}^*_{P_0}\left(Dh_{P_{12}}\right)-2u_{P_{12}}\left(D\tilde{h}^*_{P_0}\right)\right\}\\
&-3\alpha^2\left\{\tilde{h}^*_{P_0}\left(Dh_{P_{12}}\right)+2h_{P_{12}}\left(D\tilde{h}^*_{P_0}\right)\right\}\right\}+4\tilde{K}^*_0u_{P_{12}}+K_{12}\tilde{u}^*_{P_0}\\
&+(K_{10}+K^*_{10})\tilde{u}_{P_0}+D\left\{K_{12}\left(D\tilde{h}^*_{P_0}\right)-2\tilde{K}^*_0\left(Du_{P_{12}}\right)-(K_{10}+K^*_{10})\left(D\tilde{u}_{P_0}\right)\right\}\Big\}\\
&+(\theta_a-\phi_a)D\left\{2\tilde{\theta}^*_0(Du_{P_{12}})+\tilde{u}^*_{P_0}\theta_{12}+2u_{P_{12}}(D\tilde{\theta}^*_0)+\theta_{12}\left(D\tilde{u}^*_{P_0}\right)\right.\\
&\left.+\tilde{u}_{P_0}D(\theta_{10}+\theta^*_{10})\right\}-\frac{\phi_a}{\alpha^2}D\left\{\tilde{\theta}^*_0\phi_{12}+\tilde{\phi}^*_0\theta_{12}+\tilde{\theta}_0(\phi_{10}+\phi^*_{10})\right.\\
&\left.+\tilde{\phi}_0(\theta_{10}+\theta^*_{10})\right\}+2\tau h_{P_a}\left\{2h_{P_{12}}\left(D\tilde{u}^*_{P_0}\right)-\tilde{h}^*_{P_0}\left(Du_{P_{12}}\right)\right.\\
&\left.\left.+\tilde{u}^*_{P_0}\left(Dh_{P_{12}}\right)-2u_{P_{12}}\left(D\tilde{h}^*_{P_0}\right)\right\}\right]dz
\end{aligned}
$$

Integrating by parts and using the boundary conditions, finally equating LHS and RHS, we obtain an equation satisfied by the amplitude $A\,(t_s)$ as follows

$$
\frac{dA}{dt_s} = c_1A+c_2|A|^2A \tag{98}
$$

with $c_1=\frac{T_2}{T_1}$, $c_2=\frac{T_3}{T_1}$ where

$$
\begin{aligned}
T_1 = &\int_0^\delta \left[\theta_a\left(\mathcal{S}\tilde{\phi}_0-\tilde{\theta}_0\right)+\phi_a\left(1-\phi_b\right)\tilde{\theta}_0+\phi_a\left(\mathcal{C}-\theta_b\right)\tilde{\phi}_0\right.\\
&\left.+\tilde{h}_{P_0}\left\{Q\left(Du_{P_a}\right)-\tau h_{P_a}\right\}\right]dz
\end{aligned} \tag{99}
$$

$$
T_2 = \mathcal{R}_2\int_0^\delta u_{P_a}\tilde{\theta}_0 dz \tag{100}
$$

$$
\begin{aligned}
T_3 = &\ \alpha^2|A|^2A\int_0^\delta \left[u_{P_a}\frac{2Q}{\tau}\left\{D\left\{(Dh_{P_{12}})\left(D\tilde{h}^*_{P_0}\right)+\tilde{h}^*_{P_0}\left(D^2h_{P_{12}}\right)-2h_{P_{12}}\left(D^2\tilde{h}^*_{P_0}\right)\right\}\right.\right.\\
&+\tau D\left\{2h_{P_{12}}\left(D\tilde{u}^*_{P_0}\right)-\tilde{h}^*_{P_0}\left(Du_{P_{12}}\right)+\tilde{u}^*_{P_0}\left(Dh_{P_{12}}\right)-2u_{P_{12}}\left(D\tilde{h}^*_{P_0}\right)\right\}\\
&-3\alpha^2\left\{\tilde{h}^*_{P_0}\left(Dh_{P_{12}}\right)+2h_{P_{12}}\left(D\tilde{h}^*_{P_0}\right)\right\}\right\}+4\tilde{K}^*_0u_{P_{12}}+K_{12}\tilde{u}^*_{P_0}\\
&+(K_{10}+K^*_{10})\tilde{u}_{P_0}+D\left\{K_{12}\left(D\tilde{h}^*_{P_0}\right)-2\tilde{K}^*_0\left(Du_{P_{12}}\right)-(K_{10}+K^*_{10})\left(D\tilde{u}_{P_0}\right)\right\}\Big\}\\
&+(\theta_a-\phi_a)D\left\{2\tilde{\theta}^*_0(Du_{P_{12}})+\tilde{u}^*_{P_0}\theta_{12}+2u_{P_{12}}(D\tilde{\theta}^*_0)+\theta_{12}\left(D\tilde{u}^*_{P_0}\right)\right.\\
&\left.+\tilde{u}_{P_0}D(\theta_{10}+\theta^*_{10})\right\}-\frac{\phi_a}{\alpha^2}D\left\{\tilde{\theta}^*_0\phi_{12}+\tilde{\phi}^*_0\theta_{12}+\tilde{\theta}_0(\phi_{10}+\phi^*_{10})\right.\\
&\left.+\tilde{\phi}_0(\theta_{10}+\theta^*_{10})\right\}+2\tau h_{P_a}\left\{2h_{P_{12}}\left(D\tilde{u}^*_{P_0}\right)-\tilde{h}^*_{P_0}\left(Du_{P_{12}}\right)\right.\\
&\left.\left.+\tilde{u}^*_{P_0}\left(Dh_{P_{12}}\right)-2u_{P_{12}}\left(D\tilde{h}^*_{P_0}\right)\right\}\right]dz
\end{aligned} \tag{101}
$$

Multiplying equation (98) by A^* and adding this to the result obtained by multiplying the conjugate of equation (98) by A, we obtain Landau type equation

$$\frac{d\,|A|^2}{dt_s} = 2c_{1r}|A|^2 + 2c_{2r}|A|^4 \tag{102}$$

where c_{1r} and c_{2r} are real parts of c_1 and c_2 respectively. Here $-2c_{2r}$ is the Landau constant. The solution of (102) can be expressed as

$$|A|^2 = \frac{A_0^2}{\left(1 + \frac{c_{2r}}{c_{1r}} A_0^2\right) e^{-2c_{1r}t_s} - \frac{c_{2r}}{c_{1r}} A_0^2} \tag{103}$$

where A_0 is the initial value of $|A|$.

3.4. Numerical Results for Magnetic Case

To obtain our computational results for the solutions of the basic state and the linear system, we choose the parameters C = 9.0, S = 3.2 and $\theta_\infty = 0.1$ from available experimental data. The mushy layer depth is calculated from the solutions of the basic state. Computed value of mushy layer depth are 1.989359. The computational results of the present study are based on small value of Robert's number, as it is known that ([18]) the critical Rayleigh number is insensitive to smaller values of Robert's number. So we choose 0.0001 as Robert's number for all our computational work. After computing the basic state solutions, we obtain the solution of the linear system given by (73) numerically by using the fourth-order Runge-Kutta in combination of shooting method.

Figures 8 and 9 display the marginal stability curves for different values of Q and n. These results show the critical Rayleigh number increases as the permeability parameter n increases. These results indicate more efficient flow process in the presence of variable permeability which leads to smaller wavelength of the critical mode and the associated stabilizing effect.

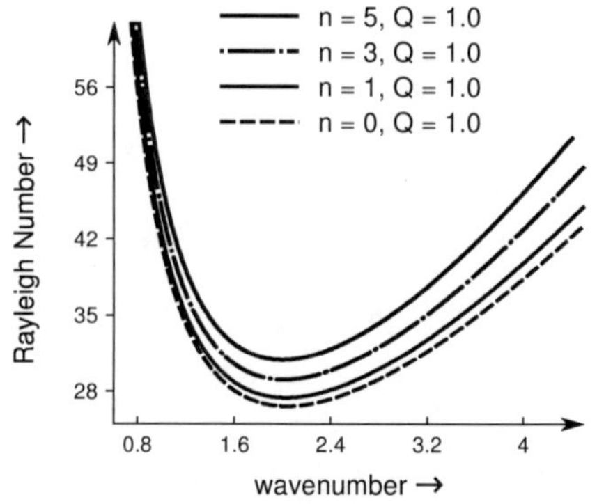

Figure 8. Marginal stability curves for different values of n for fixed $Q = 1$.

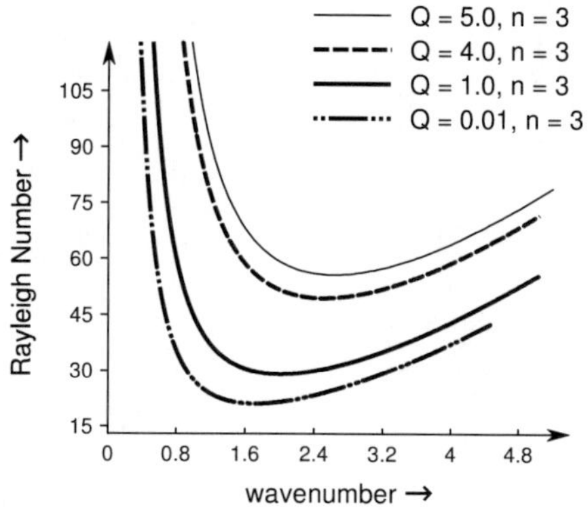

Figure 9. Marginal stability curves for different values of Q for fixed $n = 3$.

Perturbed linear components are depicted in figure 10 and figure 11 for different parameters. We take vertical axis as the axis to represent the independent variable. Figures 10 and 11 depict the dependent variable u_{P_0} for different values of t_s at $x = 0$ with $n = 3$ for $Q = 1$ and $Q = 5$ respectively. These results indicate that value of u_{P_0} reduces as time parameter t_s increases. It is clear that the maximum of u_{P_0} around the middle of mushy layer.

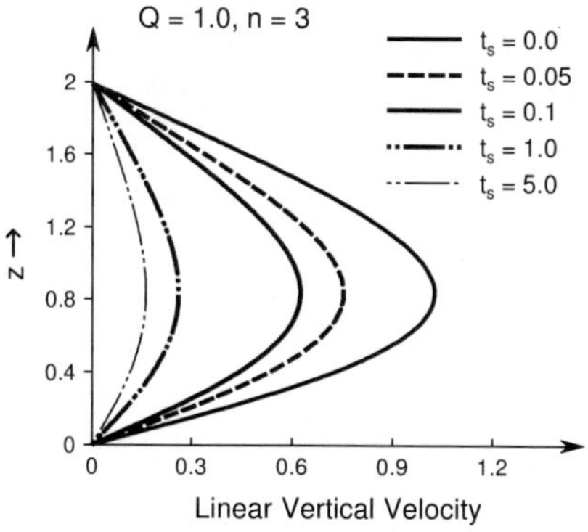

Figure 10. Perturbed linear component u_{P_0} for different values of t_s with $Q = 1$ and $n = 3$.

After obtaining the linear solutions given by (73), we solve the perturbed first-order system for the dependent variables. To obtain first-order solutions, we solve the two-α mode given by (82 - 85) and the zero-α mode given by (86-89) numerically by converting each system with seven first-order linear ordinary differential equations. Figures 12 presents the results for the perturbed first-order component for u_{P_1} at $x = 0$.

Perturbed linear solid volume fractions are shown in figures 13 and 14. The dependent variable ϕ_0 for different values of t_s at $x = 0$ with $n = 3$ is displayed in figure 13 and 14 for $Q = 1$ and $Q = 5$ respectively. Figure 15 presents the results for the perturbed first-order solid volume fraction. These results indicate that value of solid volume fraction reduces as time parameter t_s increases. These graphs suggest that as time increases there is

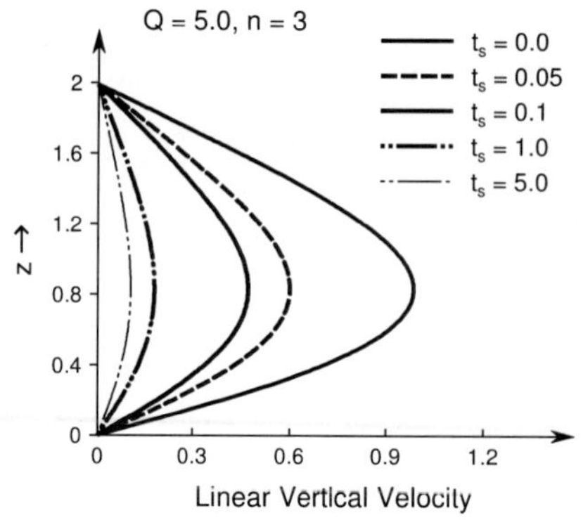

Figure 11. Perturbed linear component u_{P_0} for different values of t_s with $Q = 5$ and $n = 3$.

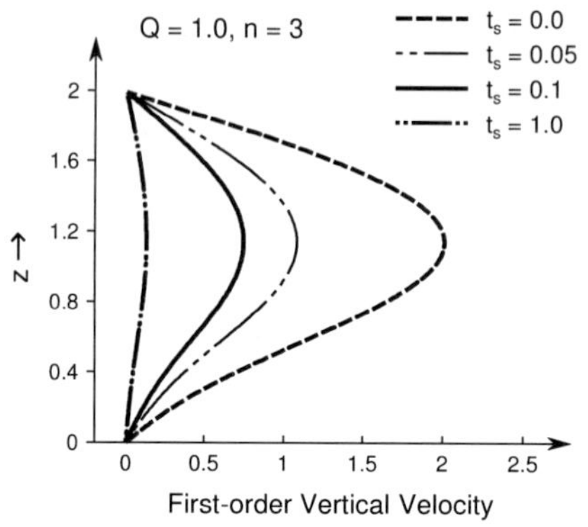

Figure 12. Perturbed first-order component u_{P_1} for different values of t_s with $Q = 1$ and $n = 3$.

less tendency for chimney formation in the linear regime, while there is more tendency for chimney formation near the lower boundary in the non-linear regime.

The perturbed first-order magnetic field components are shown in figure 16 at $x = 0$.

Now we numerically compute the solutions of the adjoint system given by (76) and use linear and first-order solutions systems to compute the coefficients of Landau type equation (102) using the expressions (99), (100), and (101) and the Simpson's rule. Table 1 presents the coefficients c_{1r}, c_{2r}, of equation (102) and asymptotic values of the amplitude $A(t_s)$ as $t_s \to \infty$ for different values of Q and n. Asymptotic value of the amplitude is independent of the initial amplitude.

It is found that the Landau constant has positive sign for all the cases we considered here. Using Landau theory, we can comment that there is a slow transition to turbulence ([1], [2]).

Variation of the amplitude function $A(t_s)$ with respect to time presented in the figure 17 for different values of n and Q. Figure 17 compares the amplitude for $n = 1$ with different values of Chandrasekhar number, Q for initial amplitude $A_0 = 1$ and it suggests that the asymptotic value of the amplitude is lower for higher values of Q.

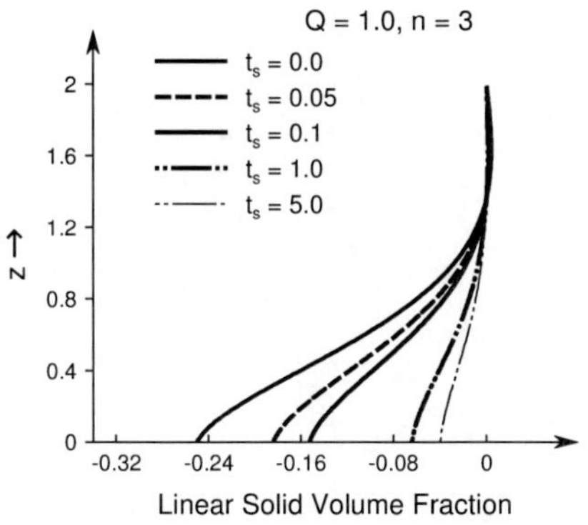

Figure 13. Perturbed linear solid volume fraction ϕ_0 for different values of t_s with $Q = 1$ and $n = 3$.

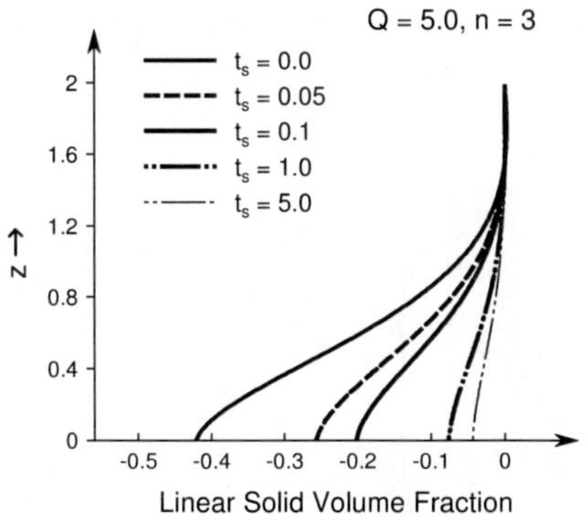

Figure 14. Perturbed linear solid volume fraction ϕ_0 for different values of t_s with $Q = 5$ and $n = 3$.

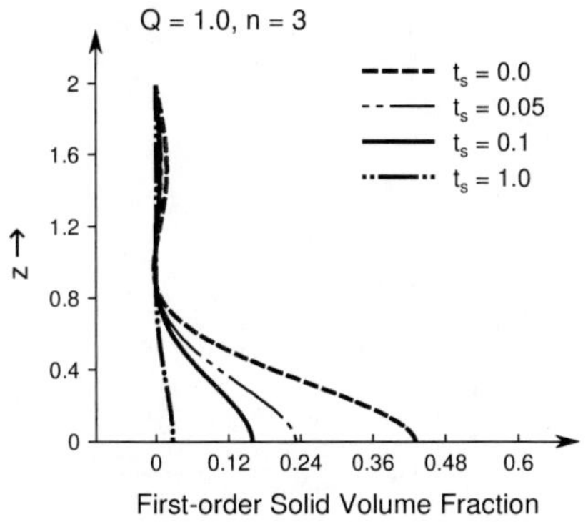

Figure 15. Perturbed first-order solid volume fraction ϕ_1 for different values of t_s with $Q = 1$ and $n = 3$.

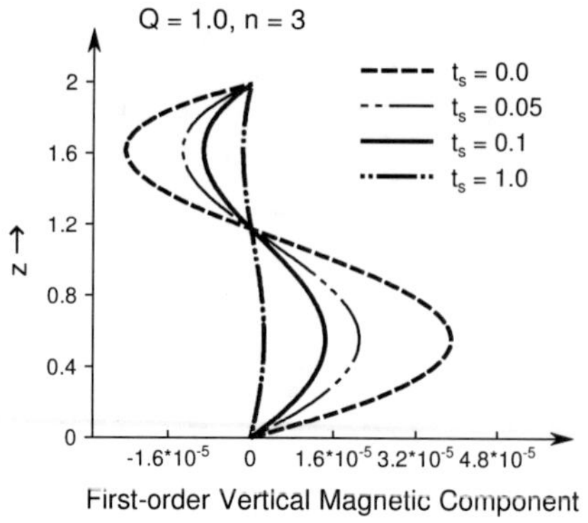

Figure 16. Perturbed first-order magnetic component for different values of t_s with $Q = 1$ and $n = 3$.

Table 1. Asymptotic values of Amplitude for different values of Q and n

Q	n	c_{1r}	c_{2r}	Aysmptotic value of $A_{t_s} = \sqrt{-\frac{c_{1r}}{c_{2r}}}$
1	0	0.2103721	-2.4267635	0.2944288
1	1	0.2040445	-7.9462916	0.1602938
1	3	0.1893518	-8.8179317	0.1465383
1	5	0.1759094	-9.3704915	0.1370135
5	0	0.1533976	-8.4130132	0.1350310
5	1	0.1492148	-17.3035545	0.0928620
5	3	0.1425197	-17.2140650	0.0909904
5	5	0.1328296	-16.0945065	0.0908466
0.1	0	0.24474508	-1.2387026	0.4445017
0.1	1	0.2344748	-4.4014843	0.2308067
0.1	3	0.2152705	-6.0878196	0.1880448
0.1	5	0.1977505	-7.2274540	0.1654116

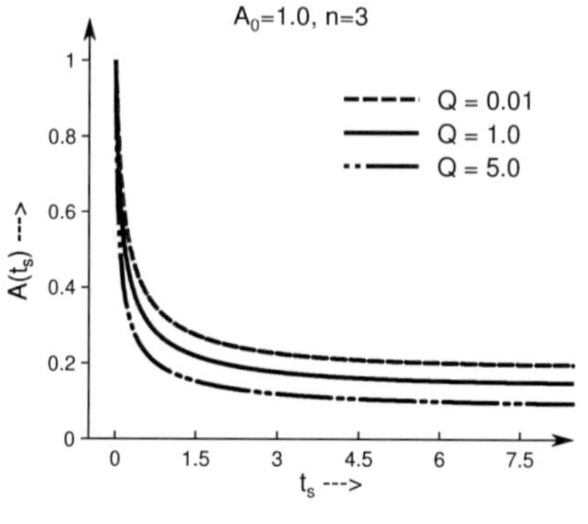

Figure 17. Amplitude $A(t_s)$ for different values of Q with $n = 3$ and $A_0 = 1$.

4. Conclusion

In this chapter, we have included a review on recent studies done by various researchers on this topic. This chapter includes an analysis on weakly nonlinear behavior of convective flow in an active mushy layer. The basic state, linear, adjoint and first-order systems are derived and these solutions are used to obtain the Landau coefficients which appear in the resulting nonlinear evolution equation. We also present computational results for marginal stability curves and other dependent variables for different permeability parameter and Chandrasekhar number. These results suggest that higher permeability parameter and Chandrasekhar number enhance the marginal stability. Numerical solutions to weakly nonlinear equations are obtained using a fourth order Runge-Kutta integration scheme via shooting technique for a set of parameter values used by experimentalists. Landau-type evolution equation is derived in terms of linear and first-order solutions by using an adjoint operator. The Landau constant is calculated for both cases: constant permeability and variable permeability. The analysis reveals that there is a steady state flow regime for large time for either a passive or an active mushy layer. However, in the case of the active layer, the amplitude of such a steady flow is smaller than the corresponding one in a passive layer. In addition, there is a slow transition to turbulence according to the Landau theory, which is insensitive to the variation of the permeability of the mushy layer. The Landau constant is positive for either case. Since based on the experimental evidence it is expected that the mushy layer with a variable permeability is more realistic, we can conclude from the present results that, at least for the parameter regime close to that for the critical one, the flow solution is not sensitive with respect to the time evolution of the flow system in its transient regime. In addition, nonlinear effect in the formation of the chimney is found to be more effective for the active layer as compared to the passive layer.

In regard to the novelty of the present results, here we make a comparison between our key results and those due to several related studies [15, 19, 24] that were based essentially on the earlier analytical modeling [13] and led to derivation of the associated amplitude equation. For the steady case where the present study was conducted, the relevant Landau constant for the associated amplitude equations derived in [15, 19] found to be positive in either constant or weakly variable permeability but larger for the non-constant permeability case. Although this result was in agreement with the present result, this by no means indicates that the previous work due to these studies provide results qualitatively similar to the present results especially for the active mushy layer case because of the following reasons. First, the previous work by those authors were based on restricted assumptions that, in particular, the solid fraction and the deviation of the permeability from a constant value were very small of the same order as the assumed small depth δ of the mushy layer ($\delta \ll 1$), and, in addition, the main permeability parameter K_1 in their work was also assumed to be small of order δ. Amberg and Homsy [13] showed that for the value of the permeability parameter K_1 of their model bigger than some small value ($K_1/C_s > 0.226$, where C_s is an order one quantity), the steady solution is possible only in subcritical regime implying a negative value of the Landau constant for the associated amplitude equation. However, such regime was not allowed to be investigated in [15] because of the assumption that $K_1 \ll 1$, and Roper, Davis and Voorhees [19] considered only constant permeability to determine the Landau constant of their amplitude equation and so their result is not applicable to the active mushy layer case. So our above explanation indicates that at least in the active mushy layer case and for the parameter values of the experiments [12] applicable to the mushy layer, the results of the present improved model for the derived amplitude equation, which is based, in particular, on an order one permeability function and led to a positive Landau constant and the subsequent results and implications, are expected to be quite different from the corresponding ones based on the earlier model [13].

Acknowledgment

Authors would like to acknowledge special thanks to the editor and the publisher.

References

[1] Landau, LD. (1944). On the problem of turbulence. *C.R. Acad. Sci. U.R.S.S* 44, 311-314.

[2] Drazin, P & Reid, WH. (1981). *Hydrodynamic Stability*. Edition. Cambridge: Cambridge University Press.

[3] Chandrasekhar, S. (1961). *Hydrodynamic and Hydromagnetic Stability*. Edition. NY, Dover Publication.

[4] Hills, R; Loper, D & Roberts, P. (1983). A thermodynamically consistent model of a mushy zone. *Q. J. Mech. Appl. Maths.*, 36, 505-539.

[5] Fowler, AC. (1985). The formation of freckles in binary alloys. *IMA. J. Appl. Math.*, 35, 159-174.

[6] Huppert, HE & Worster, MG. (1985). Dynamic solidification of a binary melt. *Nature*, 314, 703-707.

[7] Worster, MG. (1986). Solidification of an alloy from a cooled boundary. *J. Fluid Mech.*, 167, 481-501.

[8] Vives, C & Perry, C. (1987). Effects of magnetically damped convection during the controlled solidification of metals and alloys. *Int. J. Heat Mass Transfer*, 30, 479-496.

[9] Worster, MG. (1991). Natural convection in a mushy layer. *J. Fluid Mech.*, 224, 335-359.

[10] Chen, CF & Chen, CF. (1991). Experimental study directional solidification of aqueous chloride solution. *J. Fluid Mech.*, 227, 567-586.

[11] Worster, MG. (1992). Instabilities of the liquid and mushy regions during solidification of alloys. *J. Fluid Mech.*, 237, 335-359.

[12] Tait, S & Jaupart, C. (1992). The planform of compositional convection and chimney formation in a mushy layer. *Nature*, 359, 406-408.

[13] Amberg, G & Homsy, GM. (1993). Nonlinear analysis of buoyant convection in binary solidification with application to channel formation. *J. Fluid Mech.*, 252, 79-98.

[14] Emms. P & Fowler, AC. (1994). Compositional convection in the solidification of binary alloys. *J. Fluid Mech.*, 262, 111-139.

[15] Anderson, DM & Worster, MG. (1995). Weakly nonlinear analysis of convection in mushy layers during the solidification of binary alloys. *J. Fluid Mech.*, 302, 307-331.

[16] Chen, CF. (1995). Experimental study of convection in a mushy layer during directional solidification. *J. Fluid Mech.*, 293, 81-98.

[17] Riahi, DN. (1997). Effects of centrifugal and coriolis forces on chimney convection during alloy Solidification. *J. Crystal Growth*, 179, 287-296.

[18] Riahi, DN. (2000). Effects of a vertical magnetic field on chimney convection in a mushy layer. *J. Crystal Growth*, 216, 501-511.

[19] Roper, SM; Davis, SH & Voorhees, PW. (2008). An analysis of convection in a mushy layer with deformable permeable interface. *J. Fluid Mech.*, 596, 333-352.

[20] Okhuysen, BS & Riahi, DN. (2008). On weakly nonlinear convection in mushy layers during solidification of alloys. *J. Fluid Mech.*, 596, 143-167.

[21] Okhuysen, BS & Riahi, DN. (2009). Perturbation and stability analyses of flow in a mushy layer with permeable interface. *Nonlinear Analysis: Real World Applications*, 10, 3230-3239.

[22] Muddamallappa, MS; Bhatta. D & Riahi, DN. (2009). Numerical investigation on marginal stability and convection with and without magnetic field in a mushy layer. *Transport in Porous Media*, 79, 301-317.

[23] Riahi, DN. (2009). On oscillatory modes of nonlinear compositional convection in mushy layers. *Nonlinear Analysis: Real World Applications*, 10, 209-226.

[24] Guba, P & Worster, MG. (2010). Interaction between steady and oscillatory convection in mushy layers. *J. Fluid Mech.*, 645, 411-434.

[25] Bhatta, D; Muddamallappa, MS & Riahi, DN. (2010). On weakly nonlinear evolution of convective flow in a passive mushy layer. *Nonlinear Analysis: Real World Applications*, 11, 4010-4020.

[26] Bhatta, D; Riahi, DN & Muddamallappa, MS. (2012). On nonlinear evolution of convective flow in an active mushy layer. *J. Eng. Math.*, 74, 73-89.

[27] Bhatta, D; Muddamallappa, MS & Riahi, DN. *Magnetic and permeability effects on a weakly nonlinear magneto-convective flow in an active mushy layer*. Submitted, (2012).

INDEX

A

B

C

D

E

F

M

N

O

P

Q

R

S

T

U

V

W

Y